Men, Machines, and Modern Times

Men, Machines, and Modern Times

ELTING E. MORISON

The M.I.T. Press
MASSACHUSETTS INSTITUTE OF TECHNOLOGY
CAMBRIDGE, MASSACHUSETTS, AND LONDON, ENGLAND

20 19 18 17 16 15 14

ISBN 978-0-262-13025-7 (hardcover: alk. paper)

ISBN 978-0-262-63018-4 (paperback)

Library of Congress catalog card number: 66-22145
Printed in the United States of America

To the other members

Preface

In February 1950 I gave some lectures in the Athenaeum of the California Institute of Technology. I did so at the invitation of Hallett Smith, Chairman of the Division of Humanities, and Paul C. Eaton, Dean of Students. One of these lectures, which had to do with the disorder created in the United States Navy when an officer discovered a new way to fire a gun at sea, is now the first chapter of this book. It is also the source of all that follows for it started me out on the line of inquiry that is developed throughout these pages. For several years thereafter this inquiry was pursued until it seemed to me that I was reaching the point of diminishing returns, indeed the point of flat and stale repetition of earlier findings. So the subject was put aside.

In February 1963 I was invited again by the Chairman of the Division of Humanities and the Dean of Students to give three talks at the Athenaeum. In working up one of these talks I had some reflections that seemed to get me off the dead center of repetition I had been stuck on and to open up the possibility of a refreshing new direction for my earlier line of inquiry. Some of the things I said then have become the fifth chapter of this book.

Now, in February 1966, once again at the invitation of my previous hosts, I have returned for a visit to the California Institute of Technology. In writing these words in a room at the Athenaeum it appears, as they say at institutes of technology, that I am closing the loop, or in other words, that I've gone about as far as I can usefully go on this particular line of inquiry. It appears also that this is the right time and place to give my thanks to the two prime movers of this venture. In responding to their first invitation I came upon my subject. By their thought and effort, long continued, I have ever since been supplied with opportunity, means, and delightful occasions to pursue it.

I am as deep in debt to some others, too. For the last fifteen years I have been a member of a group of fourteen men who have met regularly one Friday evening a month to eat and talk together. Of the fourteen three were trained in physics, three in history, two in economics, and one each in mechanical engineering, chemistry, sociology, political sciences, psychology, and the law. More than half of them have held responsible positions in government, industry, or higher education. Collectively they have been, therefore, in fairly close touch with both the intellectual underpinnings and actual operations of an industrial society. More often than not the conversations on Friday evening have turned on subjects dealt with in these pages. This book is not to be taken as even the palest reflection of these conversations, but it would have been different, in many ways it would not have been at all, had I not known the other members.

ELTING E. MORISON

The Athenaeum
California Institute of Technology
February 1966

Contents

1

Introductory Observations, Personal and Otherwise

I was first introduced to the subject of this book thirty-four years ago in the spring of my last year in college. I had at that time gone off to my family's house in the country to complete the writing of my senior thesis. The house had been built by my great-uncle forty years earlier. He had been a civil engineer who had spent much of his professional life designing and building the bridges that carried the railroads across the Western rivers. In this house, which was itself a very complete expression of his view of things, there were still many evidences of his powerful mind and personality: a file room of his blueprints, pictures of his bridges on the walls, and, in the bookcases, copies of his engineering reports including the study he had made for the federal government (just before his death in 1901) of the feasibility of the various isthmian canal routes.

From one of these bookcases one spring evening I took down in idle curiosity a little volume he had written called

The New Epoch. It consisted of a series of papers he had delivered at various times and places in the last decade of the last century. These papers were all an elaboration of a central idea suggested to him, so he said in his preface, while reading his classmate John Fiske's *The Discovery of America.* Fiske began his work with a description of what he called the "ethnical periods among prehistoric man," epochs representing the various conditions of savagery and barbarism that led to civilization. The character of these ethnical periods was determined in large part by the tools available to the men living in the period: fire, bows and arrows, pottery, domesticated animals, ironworking, and the alphabet.

From such considerations my uncle had been led to think that he and his generation had entered a new epoch. All earlier history had been determined by the fact that the capacity of man had always been limited to his own strength and that of the men and animals he could control. But, beginning with the nineteenth century, the situation had changed. "His capacity is no longer so limited; man has now learned to *manufacture power* and with the manufacture of power a new epoch began." In this little book, in a series of essays on different subjects, the author put down his speculations on the changes — social, economic, political, and intellectual — that would be produced by man's discovery of the means to manufacture power and to distribute it to do work wherever needed. On these matters he was invariably sensible and often prescient. If in supplying possible solutions for the large problems he foresaw he betrayed, at times, the oversimplifying optimism of the nineteenth-century engineers who had solved so many problems in their own field, nevertheless his sense of the important issues was virtually unerring. The paths of inquiry he laid out in this little book were in many cases those which in later years attracted the attention of such dissimilar intelligences as Veblen, Whitehead, Schumpeter, and Wiener.

I wish I could say that all this had been clear to me at my first reading of these essays. But I was ill prepared for such speculations at the time. My reading of history in school and college had led me to believe that shape and meaning were given to society primarily by the blunders or adroit maneuvers of men who appeared in public life as generals, senators, prime ministers, foreign secretaries, kings, and ambassadors. The senior thesis on which I was working at the time was a canvass of the intricate ineptitudes both at home and abroad of the Emperor Napoleon III. One of the principal recorders of the imperial difficulties, and one of my primary sources, was Philip Guedalla, whose views of the Industrial Revolution, or the new epoch, paralleled, if they did not determine, my own. Once when he could not avoid taking railroads into account, he had said that "the change was one of those queer achievements of the nineteenth century when little men in black coats produced astonishing results whilst thinking hard all the time about something else." In the history that I had learned, from the sacking of Rome by Alaric to Bethmann-Hollweg's piece of paper, there had been small place for little men in black coats.

So the concerns of *The New Epoch* went largely unattended in the year 1932. Indeed, for some time all I retained from its pages was a statement to the effect that the power generated in a modern steamship in a single voyage across the Atlantic was more than enough to raise from the Nile and set in place every stone of the great Egyptian pyramid.

When I next read it, some seventeen or eighteen years later, I had a considerably greater interest in its subject. In the interval I had done several things that took me somewhat closer to the work of the little men in black coats when they were thinking hard. First, I had written the life of a naval officer whose whole career had been determined by the shift from sail to steam and more particularly by the increasing mechanization of the naval gunnery system. I cannot say that in the writing of this book I had been fully aware of how much that remarkable personal career had been produced simply by a change in the tools the officer used. My interest at the time had been to get at the nature and effect of a particular personality. But, at least unconsciously, I could not avoid taking in some of the more obvious implications of the influence exerted by the changes in the mechanical systems he worked with.

Following this I spent four years editing the letters of Theodore Roosevelt. I came to this task accepting the views of many in my generation that the twenty-sixth President was something of a snake-oil salesman whose hold over the citizens came from the fact that like most of them he tended, as Lincoln Steffens said, to think with his hips. Four years of labor among his papers produced an increase of understanding and a quite astonishing change of judgement. Roosevelt had a matchless sense of the nature of the democratic process as it works out in this republic, and his awareness of the place of the Presidency in that process has been equalled by only two or three of the others who have held this office. In addition, he possessed a very clear insight into the nature of the times. He was the first President to discern the meaning of what he called "the wonderful new conditions of industrial growth" and to state directly the issue of what the society would do with itself in these conditions. In the course of his years in the Presidency he put all the hard questions that were accumulating as the country steadily increased its capacity to manufacture power. He also offered interesting answers and gave it as his opinion again and again that if these answers or better alternatives were not supplied in time, the society would be shaken to pieces by the jolts and shocks produced by the continuous introduction of new energy. He was neither morbid nor depressive, but confronted by the rapid acceleration of the wonderful new conditions in his time, he had left many of the complacent assumptions of the nineteenth century — that things would proceed in good order along the whirring grooves of change — somewhat behind. He believed, in fact, that if great care were not taken in the ordering of the new energies, things would jump the rails.

Work with the letters and papers of this man was a further education. In a time when the number of little men in black coats was steadily increasing, the world was not to be held together simply by things like the Ems Dispatch or the Ostend Manifesto or the renewal of the Re-insurance Treaty. When "the strength of materials, the chemical composition of substances [and] the laws of heat and dynamic energy enter into almost every operation of modern life," some more informed and solidly based arrangements for the ordering of affairs were necessary. Education in this kind of necessity was extended when I became a member of the faculty of the Massachusetts Institute of Technology in 1946. It was at that time well on the way to becoming what it now is, one of the great intellectual centers of the new epoch. Virtually all the work done in its buildings — investigation, experiment, and the elaboration of theoretical systems — is directly related to the extended application of old forms of power and the development of means to produce new forms of power. Simply to live and work as an historian in such a place was a liberal education in the nature of the new dispensation.

Many of the people there were lineal descendants of the fellows in the black coats who produced astonishing results while thinking hard all the time about something else. But some of them were thinking, too, about the results — it was, after all, shortly after Hiroshima — and it was becoming more natural to consider not only the kind of design that would produce a mechanical system that would do more work but something of the effect of the mechanical system on the conditions of the time. Some of these people — Hawthorne, Soderberg, and Wiesner in engineering, Deutsch, Stratton, and Weisskopf in physics, Maclaurin and Millikan in economics, Bavelas in psychology, Wiener in mathematics and everything else — I got to know early on, worked and talked and, in a way, grew up with. Over the years I obtained from them at least a feeling for some of the conditions of an age that lives by the manufacture of power and some of the contemporary social implications and consequences of such manufacture. This may be an overstatement of the case, or an exaggeration of the dimension of the subject of this book. Perhaps it is fairer to say that from them I at the least began to look at the old materials in my field of history in what was for me a new way and let it go at that.

In any event, some years after I became a member of the faculty of the Institute, I thought it would be interesting to arrange some of these old materials in a new way to see what would come out of it. The thing that at the time was on many of my colleagues' minds as a new kind of inquiry — the kind of fashionable kick that so often appears to stimulate our intellectual life — was technological innovation, what Schumpeter called "the gale of creative destruction," or, more simply, what happens when you change the machinery. There are many parts of this process: why change, who changes, what design problems, what capital requirements, what institutional modifications, who gets hurt, who profits? Among the parts there is room for many of the intellectual trainings: physical scientist, engineer, economist, sociologist, psychologist, and political scientist. There might also, it seemed to me, be some place for the historian.

So I took some data with which I was thoroughly familiar, the evidence bearing on the introduction of a new system of gunnery into the United States Navy at the turn of the last century. I had used this material ten years before as part of the biography of the naval officer William S. Sims. It had then been cast in a relatively simple narrative form. This time I tried to organize it to emphasize, without undue distortion, what happened when there were changes in the mechanical systems men worked with. When the evidence was put together in this way, it appeared that there was a rather clearly defined anatomy in the whole process and a set of rather interesting connections between certain causes and effects.

The study, as thus composed, is the first essay in this book. It is also the source or beginning of all the essays that follow. In working on it, I came upon enough matters I could not resolve to my own satisfaction to start me along on further investigations. The subject since that time has remained a sort of subsidiary intellectual concern, taken up now and then from different points of view during the last fifteen years.

Much of the ensuing research and reflection has been spent on four distinct parts of the process: the condition of things at the point of origin of any mechanical change, the character of the primary agents of change, the nature of the resistance to change, and the means to facilitate general accommodation to the changes introduced. Interesting or entertaining things can be found and said on all these matters.

For instance, no man ordinarily can get out very far ahead of the state of the art, or to put it another way, he can't rise very far above the thresholds of existing knowledge. No intellectual heroism or psychic leap will take you from the development of the wheel immediately to the internal-combustion engine and the automobile. On the other hand, if the state of one art permits a considerable advance, there is not much profit in it unless the state of other related arts supports a general forward movement. The Greeks, for example, had a steam engine which remained a toy and was then forgotten because there was no way for it to do work within the technological surround, no mechanical system to hook it to. Also in Greece, the work that was done was done cheaply by slaves. There is an interesting converse to this proposition: the conditions of the technological surround, if it can inhibit change in a single part, can also foster it in a single part. Years, centuries, of experiment with windmills and waterfalls had produced by the eighteenth century a very sophisticated technology to go with these sources of energy. In some countries impressive linkages to transmit power three and four miles from the water wheels had been developed. All the elaborate machinery needed was a more effective prime mover. Since there wasn't the steam engine, it became necessary to invent it.

There are a great many other attendant propositions or tentative generalizations. It is possible, for instance, that in any very strict sense there is no such thing as an inventor or an invention. To put it another and slightly more persuasive way, the act of invention may simply be making conscious, explicit, and regular what has been done for a considerable time unconsciously or by accident. Bessemer changed his society by discovering a system for making steel in a way it had been made by accident for generations.

Also for generations before Pasteur, European mothers had been pasteurizing when they heated the milk of cows in an effort to get it as warm as their own. Indeed, almost any study of a new invention indicates that for considerable periods of time before the inventor did his work other men were also doing much the same thing.

Whether or not there are inventions it is clear that there are inventors, or at least there is a syndrome, as clearly defined as any neurosis, possessed by men who are said to invent. I once collected evidence on the lives of about thirty of these men who flourished in the nineteenth century. A surprising number turned out to be people with little formal education, who drank a good deal, who were careless with money, and who had trouble with wives or other women. This is also, I suppose, what is now called a good stereotype of the painter or poet. And it is quite probable that the inventor who is also something of an engineer is, like all great engineers, an artist and therefore shares in what is assumed to be the artistic or creative temperament. But there may be a little more to it than that. It is possible, if one sets aside the long-run social benefits, to look upon invention as a hostile act — a dislocation of existing schemes, a way of disturbing the comfortable bourgeois routines and calculations, a means of discharging the restlessness with arrangements and standards that arbitrarily limit. An Englishman who some years ago made a canvass of the lives of a good many inventors was surprised to find how many of them had worked as telegraphers. He concluded that the nature of this calling — itinerant, odd hours, episodic work loads, essentially lonely, in touch with mechanisms — supplied a kind of *rive gauche* or revolutionary underground for men not at home with standard operating procedures.

The temptation in the study of these matters historically

considered is to proceed by anecdote, for the field is rich in such materials, and then to overload the anecdote with general meaning. Nowhere is the temptation greater than in the area of resistance to change. Through the years I have collected a great many reports and episodes bearing on this subject. The one thing, for instance, which indeed every schoolboy probably does know is that after Richard Arkwright invented the spinning frame in 1769, hostile workmen tried to destroy the machines as rapidly as he could build them. There are literally thousands of such anecdotes; virtually each novelty carries with it its own tale of resistance. The ironworkers at Wyandotte burning down the first industrial laboratory in this country; the men in the vertical chain gangs that passed the buckets of water up the mine shafts in the coal fields of England who destroyed the Newcomen engines that ran the new water pumps; and so forth. Even when the new devices displaced dreadful working conditions, as in these mine shafts, the incidence and persistence of opposition are very great.

At one time the sources of this resistance appeared to me to be the single greatest matter of importance and interest in this whole process, so a good deal is said about it in the pages that follow. It still seems to me a subject deserving more attention and investigation than it has received, since without a full understanding of it no one can make much progress in dealing effectively with the last part of the process — easy general accommodation to the conditions produced by a new piece of machinery. One problem here is that men tend to continue the patterns of behavior developed in earlier conditions into the new, often quite different conditions set forth by the introduction of different mechanisms. There was a teamster I knew who had spent most of his life with horses and maintained the horses after the tractor came to take their place simply so he could take care of them and who then, after the horses had died, spent his time mending the harness for the horses that were not there. There was the mechanic at the milling machine who was kept on after a tape had been developed to guide the working of the machine but who had to be replaced after a time because he could not break himself of the habit of stopping the machine every time he was called away from the job for a short time. From anecdotes like these to modern featherbedding there is plenty of evidence that no one has yet been able to solve the problem of easy and rapid transition, for those immediately concerned, from the old to the new.

Much time is given in these pages to the examination of these topics. In one way all that is offered is a series of exercises devised to extend my own understanding of the various parts of the process. These exercises were continued over a period of six or seven years until it became clear enough that without having gotten to the bottom of the subject, I had gone as far as I could. In part this was simply a matter of diminishing returns; I was picking up information which confirmed what I already knew, which did not much advance my understanding. Much of what I had to say by way of definition and analysis had been said in the first essay on the new system of gunnery; what followed was more classification than revision and search for the more general application of a set of developed ideas than the development of ideas that were new. Furthermore, I began to realize that however useful these exercises might be in defining the process of change as it occurred in the early stages of industrialization, they would not fully serve to increase the understanding in such later stages of industrialization as we now have entered.

It will not escape notice, already no doubt has not escaped notice, that much if not most of my information is derived from the experience of the nineteenth century. It serves therefore more as the substance for illuminating parable than as confirming evidence in a contemporary situation.

For instance, the matter of invention. Throughout much of the nineteenth century inventing seemed to be the work of single men. Its occasion was, therefore, relatively haphazard and appeared often enough to be produced as much by accident as design. Today, on the other hand, we are well along into the refinement of what Whitehead called the greatest invention of the last part of the nineteenth century, the invention of the method of invention. We have pretty well left the point where the most interesting work can be done by single men working all alone. Technical knowledge is at once so specialized and so diffused that the work of one becomes in a way the work of all, which is one way of saying that the virtuosity of the inventor has on the whole given way to systematic research and development.

The refinement of method and increase of system have gone forward for another reason. The accelerating elaboration of technological structures, coupled with the advancing nicety of the fit between science and engineering, has given men much clearer contexts, physical and intellectual, to operate in. What is known in these contexts gives men something to work with; what is definable as not yet there, but necessary or desirable, gives them something to work toward. In the earlier days of more primitive contexts it was difficult to proceed with sufficient control — which sometimes helped. Goodyear was not a very impressive mind or talent. But he developed an obsessive idea that it would be nice to do something with rubber. He therefore did a great many different things with rubber, from freezing it to burying it in the ground, until one day an accident provided him with a happy thought, and he popped the substance on the stove — hence vulcanized rubber. The case of Goodyear is a little too close to turning the monkeys loose on the typewriter to produce, as a statistical probability, *Hamlet* to be of use, but it does suggest the difference between then and now. The then state of innocence may have increased the chances for the big opportunity, but it also increased the chances of failure and irrelevance in insufficient contexts. It is now possible by study of the limitations and lacunae in existing contexts to develop a rather discrete program of invention or development and to organize the intellectual means to fulfill this program. The difference between then and now is, to a considerable extent, the difference between the way in which Mr. Goodyear went about things in getting vulcanized rubber and the way in which the Bell Laboratories got to the transistor.

For most of the time I have spent in the investigation of the subject of change, concentrating as I was on the events of the nineteenth century, I did not take in even the simple fact of difference between the present and the previous condition. And the possible general implications of this change of condition I did not begin to recognize until four years ago when I was putting together the strange history of the steam vessel *Wampanoag* which appears in this volume. I cannot say that now I fully understand the meaning of these implications, but I hope I have made some progress in defining the problem.

That problem, in its two parts, seems to me to be as follows. The first part is suggested at least in the concluding pages of my great-uncle's small book. He says that "in many ways the new epoch must open as an era of destruction. It must from its very nature destroy many of the conditions which give most interest to the history of the past, and many of the traditions which people hold most dear. . . . There must be a great destruction, both in the physical and in the intellectual world, of old buildings and old boundaries and old monuments and, furthermore, of customs and ideas, systems of thought and methods of education."

Though he said that the book was not prepared in the spirit of prophecy, he went on to say two prophetic things. Remarking that the new epoch would put an end once and for all to barbarous races by the increase of education necessary to meet the new demands of the time, he discerned that "one of the greatest dangers must come from this very source, when the number of half educated people is greatest, when the world is full of people who do not know enough to recognize their limitations but know too much to follow loyally the direction of better qualified leaders." Finally he said that "the danger is that the destructive changes will come too fast, and the developments which are to take their place not fast enough. The trouble will lie in the possible gap between the two. The next two or three centuries may have periods of war, insurrection and other trials, which it would be well if the world could avoid."

He wrote these words at the turn of the century, as possible portents severely qualified by subjunctives and conditionals. His concern over such possibilities was in fact tempered by the confidence if not the complacence shared by those of his generation. Those who had spent their lives in the nineteenth century had worked with forces large enough to give them a sense for the first time in history that they were in possession of power sufficient to profoundly change the conditions of life. But the forces were never sufficiently developed to fulfill the promise they gave. The two great influences at their disposal were steam power and the large-scale production of steel. It took the better part of the century to bring these influences to a point where their impact was widely, generally felt. The power these men were dealing with, in other words, was limited; it could produce only limited change, and time was supplied to make useful accommodations to the changes made.

Today, the forces at our command, the energy in the technological structure, give shape to virtually all the conditions of life and a rapidly changing shape since we have developed the means by refining the method of invention to change the shape of the technology almost at will.

So the first part of the problem appears to be whether we can now in fact discover the means to close the gap between the changes that destroy the old, which was not bad but is not, in the new dispensation, good and useful, and the developments which are to take the place of the old, but which do not take place fast enough. Put another way, can we think of better means of accommodation than that improvised by the old teamster who mended harness after the horses were gone? And not just for the occasional man like him left momentarily technologically unemployed, but for all of us all of the time in a world where radical change is the steady state?

And the second part of the problem is like unto it. The point of invention from the earliest days of fire, iron, and the wheel has always been to give man some additional advantage over the natural environment. For much of his history the advantages obtained were, on the whole, occasional, small-scale, and, at least in the early stages, local in effect. The conditions of the natural environment, though slowly modified, continued to dominate the condition of man. Now this is changing. With our extensive knowledge and sophisticated instruments we can, in some sort, fix stars of our own in their courses and meddle with the number of the days of our years. We are well on the way, in our timeless effort to bring the natural environment under control, to replacing it by an artificial environment of our own contriving. This special environment has a structure, a set of tempos, and a series of dynamic reactions that are not always nicely scaled to human responses. The interesting question seems to be whether man, having succeeded after all these years in bringing so much of the natural environment under his control, can now manage the imposing system he has created for the specific purpose of enabling him to manage his natural environment.

As I have said, I did not begin to think about problems and questions of this kind until a short time ago. My speculations ordinarily were limited to a small sector of the general subject, a sector of perhaps diminishing significance within the whole. I worked slowly out from the events surrounding a change in a six-inch gun mount, through a new development in the dairy industry, to a tentative investigation of the use of computers. Evidence obtained in these studies, as already noted, permitted reflections on such matters as the process and immediate effects of change in specific pieces of machinery, such things as the nature of individual responses to change, the costs of accommodation to change, the impact of change on institutional structure, the means to keep what is now called creativity in formal institutions, and so forth. Such thoughts as I have had on these subjects appear in the essays that follow. Pursuing them over the years I was led to a concern with the larger topic of the character of the new epoch. In the final chapter of this book such tentative conclusions as I have reached on this topic are presented.

2

Gunfire at Sea: A Case Study of Innovation

In the early days of the last war when armaments of all kinds were in short supply, the British, I am told, made use of a venerable field piece that had come down to them from previous generations.* The honorable past of this light artillery stretched back, in fact, to the Boer War. In the days of uncertainty after the fall of France, these guns, hitched to trucks, served as useful mobile units in the coast defense. But it was felt that the rapidity of fire could be increased. A time-motion expert was, therefore, called in to suggest ways to simplify the firing procedures. He watched one of the gun crews of five men at practice in the field for some time. Puzzled by certain aspects of the procedures, he took some slow-motion pictures of the

* This essay was delivered as one of three lectures at the California Institute of Technology in 1950. It has been reprinted in various truncated forms a good many times since. This is the first time it has appeared as it was originally written.

soldiers performing the loading, aiming, and firing routines.

When he ran these pictures over once or twice, he noticed something that appeared odd to him. A moment before the firing, two members of the gun crew ceased all activity and came to attention for a three-second interval extending throughout the discharge of the gun. He summoned an old colonel of artillery, showed him the pictures, and pointed out this strange behavior. What, he asked the colonel, did it mean. The colonel, too, was puzzled. He asked to see the pictures again. "Ah," he said when the performance was over, "I have it. They are holding the horses."

This story, true or not, and I am told it is true, suggests nicely the pain with which the human being accommodates himself to changing conditions. The tendency is apparently involuntary and immediate to protect oneself against the shock of change by continuing in the presence of altered situations the familiar habits, however incongruous, of the past.

Yet, if human beings are attached to the known, to the realm of things as they are, they also, regrettably for their peace of mind, are incessantly attracted to the unknown and things as they might be. As Ecclesiastes glumly pointed out, men persist in disordering their settled ways and beliefs by seeking out many inventions.

The point is obvious. Change has always been a constant in human affairs; today, indeed, it is one of the determining characteristics of our civilization. In our relatively shapeless social organization, the shifts from station to station are fast and easy. More important for our immediate purpose, America is fundamentally an industrial society in a time of tremendous technological development. We are thus constantly presented with new devices or new forms of power that in their refinement and extension continually bombard the fixed structure of our habits of mind and behavior. Under such conditions, our salvation, or at least our peace of mind, appears to depend upon how successfully we can in the future become what has been called in an excellent phrase a completely "adaptive society."

It is interesting, in view of all this, that so little investigation, relatively, has been made of the process of change and human responses to it. Recently, psychologists, sociologists, cultural anthropologists, and economists have addressed themselves to the subject with suggestive results. But we are still far from a full understanding of the process and still further from knowing how we can set about simplifying and assisting an individual's or a group's accommodation to new machines or new ideas.

With these things in mind, I thought it might be interesting and perhaps useful to examine historically a changing situation within a society; to see if from this examination we can discover how the new machines or ideas that introduced the changing situation developed; to see who introduces them, who resists them, what points of friction or tension in the social structure are produced by the innovation, and perhaps why they are produced and what, if anything, may be done about it. For this case study the introduction of continuous-aim firing in the United States Navy has been selected. The system, first devised by an English officer in 1898, was introduced in our Navy in the years 1900 to 1902.

I have chosen to study this episode for two reasons. First, a navy is not unlike a society that has been placed under laboratory conditions. Its dimensions are severely limited; it is beautifully ordered and articulated; it is relatively isolated from random influences. For these reasons the impact of change can be clearly discerned, the resulting dislocations in the structure easily discovered and marked out. In the second place, the development of continuous-aim firing rests upon mechanical devices. It therefore presents for study a concrete, durable situation. It is not like many other innovating reagents — a Manichean heresy, or Marxism, or the views of Sigmund Freud — that can be shoved and hauled out of shape by contending forces or conflicting prejudices. At all times we know exactly what continuous-aim firing really is. It will be well now to describe, as briefly as possible, what it really is. This will involve a short investigation of certain technical matters. I will not apologize, as I have been told I ought to do, for this preoccupation with how a naval gun is fired. For one thing, all that follows is understandable only if one understands how the gun goes off. For another thing, a knowledge of the underlying physical considerations may give a kind of elegance to the succeeding investigation of social implications. And now to the gun and the gunfire.

The governing fact in gunfire at sea is that the gun is mounted on an unstable platform, a rolling ship. This constant motion obviously complicates the problem of holding a steady aim. Before 1898 this problem was solved in the following elementary fashion. A gun pointer estimated the range of the target, ordinarily in the nineties about 1600 yards. He then raised the gun barrel to give the gun the elevation to carry the shell to the target at the estimated range. This elevating process was accomplished by turning a small wheel on the gun mount that operated the elevating gears. With the gun thus fixed for range, the gun pointer peered through open sights, not unlike those on a small rifle, and waited until the roll of the ship brought the sights on the target. He then pressed the firing button that discharged the gun. There were by 1898, on some naval guns, telescope sights which naturally greatly enlarged the image of the target for the gun pointer. But these sights were rarely used by gun pointers. They were lashed securely to the gun barrel, and, recoiling with the barrel, jammed back against the unwary pointer's eye. Therefore, when used at all, they were used only to take an initial sight for purposes of estimating the range before the gun was fired.

Notice now two things about the process. First of all, the rapidity of fire was controlled by the rolling period of the ship. Pointers had to wait for the one moment in the roll when the sights were brought on the target. Notice also this: There is in every pointer what is called a "firing interval"—that is, the time lag between his impulse to fire the gun and the translation of this impulse into the act of pressing the firing button. A pointer, because of this reaction time, could not wait to fire the gun until the exact moment when the roll of the ship brought the sights onto the target; he had to will to fire a little before, while the sights were off the target. Since the firing interval was an individual matter, varying obviously from man to man, each pointer had to estimate from long practice his own interval and compensate for it accordingly.

These things, together with others we need not here investigate, conspired to make gunfire at sea relatively uncertain and ineffective. The pointer, on a moving platform, estimating range and firing interval, shooting while his sight was off the target, became in a sense an individual artist.

In 1898, many of the uncertainties were removed from the process and the position of the gun pointer radically altered by the introduction of continuous-aim firing. The major change was that which enabled the gun pointer to keep his sight and gun barrel on the target throughout the roll of the ship. This was accomplished by altering the gear ratio in the elevating gear to permit a pointer to compensate for the roll of the vessel by rapidly elevating and depressing the gun. From this change another followed. With the possibility of maintaining the gun always on the target, the desirability of improved sights became immediately apparent. The advantages of the telescope sight as opposed to the open sight were for the first time fully realized. But the existing telescope sight, it will be recalled, moved with the recoil of the gun and jammed back against the eye of the gunner. To correct this, the sight was mounted on a sleeve that permitted the gun barrel to recoil through it without moving the telescope.

These two improvements in elevating gear and sighting eliminated the major uncertainties in gunfire at sea and greatly increased the possibilities of both accurate and rapid fire.

You must take my word for it, since the time allowed is small, that this changed naval gunnery from an art to a science, and that gunnery accuracy in the British and our Navy increased, as one student said, 3000% in six years. This does not mean much except to suggest a great increase in accuracy. The following comparative figures may mean a little more. In 1899 five ships of the North Atlantic Squadron fired five minutes each at a lightship hulk at the conventional range of 1600 yards. After twenty-five minutes of banging away, two hits had been made on the sails of the elderly vessel. Six years later one naval gunner made fifteen hits in one minute at a target 75 by 25 feet at the same range — 1600 yards; half of them hit in a bull's-eye 50 inches square.

Now with the instruments (the gun, elevating gear, and telescope), the method, and the results of continuous-aim firing in mind, let us turn to the subject of major interest:

how was the idea, obviously so simple an idea, of continuous-aim firing developed, who introduced it into the United States Navy, and what was its reception?

The idea was the product of the fertile mind of the English officer Admiral Sir Percy Scott. He arrived at it in this way while, in 1898, he was the captain of H.M.S. *Scylla.* For the previous two or three years he had given much thought independently and almost alone in the British Navy to means of improving gunnery. One rough day, when the ship, at target practice, was pitching and rolling violently, he walked up and down the gun deck watching his gun crews. Because of the heavy weather, they were making very bad scores. Scott noticed, however, that one pointer was appreciably more accurate than the rest. He watched this man with care, and saw, after a time, that he was unconsciously working his elevating gear back and forth in a partially successful effort to compensate for the roll of the vessel. It flashed through Scott's mind at that moment that here was the sovereign remedy for the problem of inaccurate fire. What one man could do partially and unconsciously perhaps all men could be trained to do consciously and completely.

Acting on this assumption, he did three things. First, in all the guns of the *Scylla,* he changed the gear ratio in the elevating gear, previously used only to set the gun in fixed position for range, so that a gunner could easily elevate and depress the gun to follow a target throughout the roll. Second, he rerigged his telescopes so that they would not be influenced by the recoil of the gun. Third, he rigged a small target at the mouth of the gun, which was moved up and down by a crank to simulate a moving target. By following this target as it moved and firing at it with a subcaliber rifle rigged in the breech of the gun, the pointer could practice every day. Thus equipped, the ship became a training ground for gunners. Where before the good pointer was an individual artist, pointers now became trained technicians, fairly uniform in their capacity to shoot. The effect was immediately felt. Within a year the *Scylla* established records that were remarkable.

At this point I should like to stop a minute to notice several things directly related to, and involved in, the process of innovation. To begin with, the personality of the innovator. I wish there were time to say a good deal about Admiral Sir Percy Scott. He was a wonderful man. Three small bits of evidence must here suffice, however. First, he had a certain mechanical ingenuity. Second, his personal life was shot through with frustration and bitterness. There was a divorce and a quarrel with that ambitious officer Lord Charles Beresford, the sounds of which, Scott liked to recall, penetrated to the last outposts of empire. Finally, he possessed, like Swift, a savage indignation directed ordinarily at the inelastic intelligence of all constituted authority, especially the British Admiralty.

There are other points worth mention here. Notice first that Scott was not responsible for the invention of the basic instruments that made the reform in gunnery possible. This reform rested upon the gun itself, which as a rifle had been in existence on ships for at least forty years; the elevating gear, which had been, in the form Scott found it, a part of the rifled gun from the beginning; and the telescope sight, which had been on shipboard at least eight years. Scott's contribution was to bring these three elements appropriately modified into a combination that made continuous-aim firing possible for the first time. Notice also that he was allowed to bring these elements into combination by accident, by watching the unconscious action of a gun pointer endeavoring through the operation of his elevating gear to correct partially for the roll of his vessel. Scott, as we have seen, had been interested in gunnery; he had thought about ways to increase accuracy by practice and improvement of existing machinery; but able as he was, he had not been able to produce on his own initiative and by his own thinking the essential idea and modify instruments to fit his purpose. Notice here, finally, the intricate interaction of chance, the intellectual climate, and Scott's mind. Fortune (in this case, the unaware gun pointer) indeed favors the prepared mind, but even fortune and the prepared mind need a favorable environment before they can conspire to produce sudden change. No intelligence can proceed very far above the threshold of existing data or the binding combinations of existing data.

All these elements that enter into what may be called "original thinking" interest me as a teacher. Deeply rooted in the pedagogical mind often enough is a sterile infatuation with "inert ideas"; there is thus always present in the profession the tendency to be diverted from the *process* by which these ideas, or indeed any ideas, are really produced. I well remember with what contempt a class of mine which was reading Leonardo da Vinci's *Notebooks* dismissed the author because he appeared to know no more mechanics than, as one wit in the class observed, a Vermont Republican farmer of the present day. This is perhaps the expected result produced by a method of instruction that too frequently implies that the great generalizations were the result, on the one hand, of chance — an apple falling in an orchard or a teapot boiling on the hearth — or, on the other hand, of some towering intelligence proceeding in isolation inexorably toward some prefigured idea, like evolution, for example.

This process by which new concepts appear, the interaction of fortune, intellectual climate, and the prepared imaginative mind, is an interesting subject for examination offered by any case study of innovation. It was a subject as Dr. Walter Cannon pointed out, that momentarily engaged the attention of Horace Walpole, whose lissome intelligence glided over the surface of so many ideas. In reflecting upon the part played by chance in the development of new concepts, he recalled the story of the three princes of Serendip who set out to find some interesting object on a journey through their realm. They did not find the particular object of their search, but along the way they discovered many new things simply because they were looking for *something*. Walpole believed this intellectual method ought to be given a name, in honor of the founders, serendipity; and serendipity certainly exerts a considerable influence in what we call original thinking. There is an element of serendipity, for example, in Scott's chance discovery of continuous-aim firing in that he was, and had been, looking for some means to improve his target practice and stumbled upon a solution by observation that had never entered his head.

Serendipity, while recognizing the prepared mind, does tend to emphasize the role of chance in intellectual discovery. Its effect may be balanced by an anecdote that suggests the contribution of the adequately prepared mind. There has recently been much posthaste and romage in the land over the question of whether there really was a Renaissance. A scholar has recently argued in print that since the Middle Ages actually possessed many of the instruments and pieces of equipment associated with the Renaissance, the Renaissance could be said to exist as a defined period only in the mind of the historians such as Burckhardt. This view was entertainingly rebutted by the historian of art Panofsky, who pointed out that although Robert Grosseteste indeed did have a very rudimentary telescope, he used it to examine stalks of grain in a field down the street. Galileo, a Renaissance intelligence, pointed his telescope at the sky.

Here Panofsky is only saying in a provocative way that change and intellectual advance are the products of well-trained and well-stored inquisitive minds, minds that relieve us of "the terrible burden of inert ideas by throwing them into a new combination." Educators, nimble in the task of pouring the old wine of our heritage into the empty vessels that appear before them, might give thought to how to develop such independent, inquisitive minds.

But I have been off on a private venture of my own. Now to return to the story, the introduction of continuous-aim firing. In 1900 Percy Scott went out to the China Station as commanding officer of H.M.S. *Terrible*. In that ship he continued his training methods and his spectacular successes in naval gunnery. On the China Station he met up with an American junior officer, William S. Sims. Sims had little of the mechanical ingenuity of Percy Scott, but the two were drawn together by temperamental similarities that are worth noticing here. Sims had the same intolerance for what is called spit and polish and the same contempt for bureaucratic inertia as his British brother officer. He had for some years been concerned, as had Scott, with what he took to be the inefficiency of his own Navy. Just before he met Scott, for example, he had shipped out to China in the brand new pride of the fleet, the battleship *Kentucky*. After careful investigation and reflection he had informed his superiors in Washington that she was "not a battleship at all — but a crime against the white race." The spirit with which he pushed forward his efforts to reform the naval service can best be stated in his own words to a brother officer: "I am perfectly willing that those holding views differing from mine should continue to live, but with every fibre of my being I loathe indirection and shiftiness, and where it occurs in high place, and is used to save face at the expense of the vital interests of our great service (in which silly people place such a child-like trust), I want that man's blood and I will have it no matter what it costs me personally."

From Scott in 1900 Sims learned all there was to know about continuous-aim firing. He modified, with the Englishman's active assistance, the gear on his own ship and tried out the new system. After a few months' training, his experimental batteries began making remarkable records at target practice. Sure of the usefulness of his gunnery methods, Sims then turned to the task of educating the Navy at large. In thirteen great official reports he documented the case for continuous-aim firing, supporting his arguments at every turn with a mass of factual data. Over a period of two years, he reiterated three principal points: first, he continually cited the records established by Scott's ships, the *Scylla* and the *Terrible,* and supported these with the accumulating data from his own tests on an American ship; second, he described the mechanisms used and the training procedures instituted by Scott and himself to obtain these records; third, he explained that our own mechanisms were not generally adequate without modification to meet the demands placed on them by continuous-aim firing. Our elevating gear, useful to raise or lower a gun slowly to fix it in position for the proper range, did not always work easily and rapidly enough to enable a gunner to follow a target with his gun throughout the roll of the ship. Sims also explained that such few telescope sights as there were on board our ships were useless. Their cross wires were so thick or coarse they obscured the target, and the sights had been attached to the gun in such a way that the recoil system of the gun plunged the eyepiece against the eye of the gun pointer.

This was the substance not only of the first but of all the succeeding reports written on the subject of gunnery from the China Station. It will be interesting to see what response these met with in Washington. The response falls roughly into three easily identifiable stages.

First stage: At first, there was no response. Sims had directed his comments to the Bureau of Ordnance and the Bureau of Navigation; in both bureaus there was dead silence. The thing — claims and records of continuous-aim firing — was not credible. The reports were simply filed away and forgotten. Some indeed, it was later discovered to Sims's delight, were half-eaten-away by cockroaches.

Second stage: It is never pleasant for any man's best work to be left unnoticed by superiors, and it was an unpleasantness that Sims suffered extremely ill. In his later reports, beside the accumulating data he used to clinch his argument, he changed his tone. He used deliberately shocking language because, as he said, "They were furious at my first papers and stowed them away. I therefore made up my mind I would give these later papers such a form that they would be dangerous documents to leave neglected in the files." To another friend he added, "I want scalps or nothing and if I can't have 'em I won't play."

Besides altering his tone, he took another step to be sure his views would receive attention. He sent copies of his reports to other officers in the fleet. Aware as a result that Sims's gunnery claims were being circulated and talked about, the men in Washington were then stirred to action. They responded, notably through the Chief of the Bureau of Ordnance, who had general charge of the equipment used in gunnery practice, as follows: (1) our equipment was in general as good as the British; (2) since our equipment was as good, the trouble must be with the men, but the gun pointer and the training of gun pointers were the responsibility of the officers on the ships; and most significant (3) continuous-aim firing was impossible. Experiments had revealed that five men at work on the elevating gear of a six-inch gun could not produce the power necessary to compensate for a roll of five degrees in ten seconds. These experiments and calculations demonstrated beyond peradventure or doubt that Scott's system of gunfire was not possible.

This was the second stage — the attempt to meet Sims's claims by logical, rational rebuttal. Only one difficulty is discoverable in these arguments; they were wrong at important points. To begin with, while there was little difference between the standard British equipment and the standard American equipment, the instruments on Scott's two ships, the *Scylla* and the *Terrible,* were far better than the standard equipment on our ships. Second, all the men could not be trained in continuous-aim firing until equipment was improved throughout the fleet. Third, the experiments with the elevating gear had been ingeniously contrived at the Washington Navy Yard — on solid ground. It had, therefore, been possible to dispense in the Bureau of Ordnance calculation with Newton's first law of motion, which naturally operated at sea to assist the gunner in elevating or depressing a gun mounted on a moving ship. Another difficulty was of course that continuous-aim firing was in use on Scott's and some of our own ships at the time the Chief of the Bureau of Ordnance was writing that it was a mathematical impossibility. In every way I find this second stage, the apparent resort to reason, the most entertaining and instructive in our investigation of the responses to innovation.

Third stage: The rational period in the counterpoint between Sims and the Washington men was soon passed. It was followed by the third stage, that of name-calling —

the *argumentum ad hominem*. Sims, of course, by the high temperature he was running and by his calculated overstatement, invited this. He was told in official endorsements on his reports that there were others quite as sincere and loyal as he and far less difficult; he was dismissed as a crackbrained egotist; he was called a deliberate falsifier of evidence.

The rising opposition and the character of the opposition were not calculated to discourage further efforts by Sims. It convinced him that he was being attacked by shifty, dishonest men who were the victims, as he said, of insufferable conceit and ignorance. He made up his mind, therefore, that he was prepared to go to any extent to obtain the "scalps" and the "blood" he was after. Accordingly, he, a lieutenant, took the extraordinary step of writing the President of the United States, Theodore Roosevelt, to inform him of the remarkable records of Scott's ships, of the inadequacy of our own gunnery routines and records, and of the refusal of the Navy Department to act. Roosevelt, who always liked to respond to such appeals when he conveniently could, brought Sims back from China late in 1902 and installed him as Inspector of Target Practice, a post the naval officer held throughout the remaining six years of the Administration. And when he left, after many spirited encounters we cannot here investigate, he was universally acclaimed as "the man who taught us how to shoot."

With this sequence of events (the chronological account of the innovation of continuous-aim firing) in mind, it is possible now to examine the evidence to see what light it may throw on our present interest: the origins of and responses to change in a society.

First, the origins. We have already analyzed briefly the origins of the idea. We have seen how Scott arrived at his notion. We must now ask ourselves, I think, why Sims so actively sought, almost alone among his brother officers, to introduce the idea into his service. It is particularly interesting here to notice again that neither Scott nor Sims invented the instruments on which the innovation rested. They did not urge their proposal, as might be expected, because of pride in the instruments of their own design. The telescope sight had first been placed on shipboard in 1892 by Bradley Fiske, an officer of great inventive capacity. In that year Fiske had even sketched out on paper the vague possibility of continuous-aim firing, but his sight was condemned by his commanding officer, Robley D. Evans, as of no use. In 1892 no one but Fiske in the Navy knew what to do with a telescope sight any more than Grosseteste had known in his time what to do with a telescope. And Fiske, instead of fighting for his telescope, turned his attention to a range finder. But six years later Sims, following the tracks of his brother officer, took over and became the engineer of the revolution. I would suggest, with some reservations, this explanation: Fiske, as an inventor, took his pleasure in great part from the design of the device. He lacked not so much the energy as the overriding sense of social necessity that would have enabled him to *force* revolutionary ideas on the service. Sims possessed this sense. In Fiske, who showed rare courage and integrity in other professional matters not intimately connected with the introduction of new weapons of his own design, we may here find the familiar plight of the engineer who often enough must watch the products of his ingenuity organized and promoted by other men. These other promotional men when they appear in the world of commerce are called entrepreneurs. In the world of ideas they are still entrepreneurs. Sims was one, a middle-aged man caught in the periphery (as a lieutenant) of the intricate webbing of a precisely organized society. Rank, the exact definition and limitation of a man's capacity at any given moment in his career, prevented Sims from discharging all his exploding energies into the purely routine channels of the peacetime Navy. At the height of his powers he was a junior officer standing watches on a ship cruising aimlessly in friendly foreign waters. The remarkable changes in systems of gunfire to which Scott introduced him gave him the opportunity to expend his energies quite legitimately against the encrusted hierarchy of his society. He was moved, it seems to me, in part by his genuine desire to improve his own profession but also in part by rebellion against tedium, against inefficiency from on high, and against the artificial limitations placed on his actions by the social structure, in his case, junior rank.

Now having briefly investigated the origins of the change, let us examine the reasons for what must be considered the weird response we have observed to this proposed change. Why this deeply rooted, aggressive, persistent hostility from Washington that was only broken up by the interference of Theodore Roosevelt? Here was a reform that greatly and demonstrably increased the fighting effectiveness of a service that maintains itself almost exclusively to fight. Why then this refusal to accept so carefully documented a case, a case proved incontestably by records and experience? Why should virtually all the rulers of a society so resolutely seek to reject a change that so markedly improved its chances for survival in any contest with competing societies? There are the obvious reasons that will occur to all of you — the source of the proposed reform was an obscure, junior officer 8000 miles away; he was, and this is a significant factor, criticizing gear and machinery designed by the very men in the bureaus to whom he was sending his criticisms. And furthermore, Sims was seeking to introduce what he claimed were improvements in a field where improvements appeared unnecessary. Superiority in war, as in other things, is a relative matter, and the Spanish-American War had been won by the old system of gunnery. Therefore, it was superior even though of the 9500 shots fired at various but close ranges, only 121 had found their mark.

These are the more obvious, and I think secondary or supporting, sources of opposition to Sims's proposed reforms. A less obvious cause appears by far the most important one. It has to do with the fact that the Navy is not only an armed force; it is a society. Men spend their whole lives in it and tend to find the definition of their whole being within it. In the forty years following the Civil War, this society had been forced to accommodate itself to a series of technological changes — the steam turbine, the electric motor, the rifled shell of great explosive power, case-hardened steel armor, and all the rest of it. These changes wrought extraordinary changes in ship design, and, therefore, in the concepts of how ships were to be used; that is, in fleet tactics, and even in naval strategy. The Navy of this period is a paradise for the historian or sociologist in search of evidence bearing on a society's responses to change.

To these numerous innovations, producing as they did a spreading disorder throughout a service with heavy commitments to formal organization, the Navy responded with grudging pain. For example, sails were continued on our first-line ships long after they ceased to serve a useful purpose mechanically, but like the holding of the horses that no longer hauled the British field pieces, they assisted officers over the imposing hurdles of change. To a man raised in sail, a sail on an armored cruiser propelled through the water at 14 knots by a steam turbine was a cheering sight to see.

This reluctance to change with changing conditions was not limited to the blunter minds and less resilient imaginations in the service. As clear and untrammeled an intelligence as Alfred Thayer Mahan, a prophetic spirit in the realm of strategy, where he was unfettered by personal attachments of any kind, was occasionally at the mercy of the past. In 1906 he opposed the construction of battleships with single-caliber main batteries — that is, the modern battleship — because, he argued, such vessels would fight only at great ranges. These ranges would create in the sailor what Mahan felicitously called "the indisposition to close." They would thus undermine the physical and moral courage of a commander. They would, in other words, destroy the doctrine and the spirit, formulated by Nelson a century before, that no captain could go very far wrong who laid his ship alongside an enemy. The fourteen-inch rifle, which could place a shell upon a possible target six miles away, had long ago annihilated the Nelsonian doctrine. Mahan, of course, knew and recognized this fact; he was, as a man raised in sail, reluctant only to accept its full meaning, which was not that men were no longer brave, but that 100 years after the battle of the Nile they had to reveal their bravery in a different way.

Now the question still is, why this blind reaction to technological change, observed in the continuation of sail or in Mahan's contentions or in the opposition to continuous-aim firing? It is wrong to assume, as it is frequently assumed by civilians, that it springs exclusively from some causeless Bourbon distemper that invades the military mind. There is a sounder and more attractive base. The opposition, where it occurs, of the soldier and the sailor to such change springs from the normal human instinct to protect oneself, and more especially, one's way of life. Military organizations are societies built around and upon the prevailing weapons systems. Intuitively and quite correctly the military man feels that a change in weapon portends a change in the arrangements of his society. Think of it this way. Since the time that the memory of man runneth not to the contrary, the naval society has been built upon the surface vessel. Daily routines, habits of mind, social organization, physical accommodations, conventions, rituals, spiritual allegiances have been conditioned by the essential fact of the ship. What then happens to your society if the ship is displaced as the principal element by such a radically different weapon as the plane? The mores and structure of the society are immediately placed in jeopardy. They may, in fact, be wholly destroyed. It was the witty cliché of the twenties that those naval officers who persisted in defending the battleship against the apparently superior claims of the carrier did so because the battleship was a more comfortable home. What, from one point of view, is a better argument? There is, as everyone knows, no place like home. Who has ever wanted to see the old place brought under the hammer by hostile forces whether they hold a mortgage or inhabit a flying machine?

This sentiment would appear to account in large part for the opposition to Sims; it was the product of an instinctive protective feeling, even if the reasons for this feeling were not overt or recognized. The years after 1902 proved how right, in their terms, the opposition was. From changes in gunnery flowed an extraordinary complex of changes: in shipboard routines, ship design, and fleet tactics. There was, too, a social change. In the days when gunnery was taken lightly, the gunnery officer was taken lightly. After 1903, he became one of the most significant and powerful members of a ship's company, and this shift of emphasis naturally was shortly reflected in promotion lists. Each one of these changes provoked a dislocation in the naval society, and with man's troubled foresight and natural indisposition to break up classic forms, the men in Washington withstood the Sims onslaught as long as they could. It is very significant that they withstood it until an agent from outside, outside and above, who was not clearly identified with the naval society, entered to force change.

This agent, the President of the United States, might reasonably and legitimately claim the credit for restoring our gunnery efficiency. But this restoration by *force majeure* was brought about at great cost to the service and men involved. Bitternesses, suspicions, wounds were made that it was impossible to conceal and were, in fact, never healed.

Now this entire episode may be summed up in five separate points:

1. The essential idea for change occurred in part by chance but in an environment that contained all the essential elements for change and to a mind prepared to recognize the possibility of change.

2. The basic elements, the gun, gear, and sight, were put in the environment by other men, men interested in designing machinery to serve different purposes or simply interested in the instruments themselves.

3. These elements were brought into successful combination by minds not interested in the instruments for themselves but in what they could do with them. These minds were, to be sure, interested in good gunnery, overtly and consciously. They may also, not so consciously, have

been interested in the implied revolt that is present in the support of all change. Their temperaments and careers indeed support this view. From gunnery, Sims went on to attack ship designs, existing fleet tactics, and methods of promotion. He lived and died, as the service said, a stormy petrel, a man always on the attack against higher authority, a rebellious spirit; a rebel, fighting in excellent causes, but a rebel still who seems increasingly to have identified himself with the act of revolt against constituted authority.

4. He and his colleagues were opposed on this occasion by men who were apparently moved by three considerations: honest disbelief in the dramatic but substantiated claims of the new process, protection of the existing devices and instruments with which they identified themselves, and maintenance of the existing society with which they were identified.

5. The deadlock between those who sought change and those who sought to retain things as they were was broken only by an appeal to superior force, a force removed from and unidentified with the mores, conventions, devices of the society. This seems to me a very important point. The naval society in 1900 broke down in its effort to accommodate itself to a new situation. The appeal to Roosevelt is documentation for Mahan's great generalization that no military service should or can undertake to reform itself. It must seek assistance from outside.

Now with these five summary points in mind, it may be possible to seek, as suggested at the outset, a few larger implications from this story. What, if anything, may it suggest about the general process by which any society attempts to meet changing conditions?

There is, to begin with, a disturbing inference half-concealed in Mahan's statement that no military organization can reform itself. Certainly civilians would agree with this. We all know now that war and the preparation for war are too important, as Clemenceau said, to be left to the generals. But as I have said before, military organizations are really societies, more rigidly structured, more highly integrated, than most communities, but still societies. What then if we make this phrase to read, "No society can reform itself"? Is the process of adaptation to change, for example, too important to be left to human beings? This is a discouraging thought, and historically there is some cause to be discouraged. Societies have not been very successful in reforming themselves, accommodating to change, without pain and conflict.

This is a subject to which we may well address ourselves. Our society especially is built, as I have said, just as surely upon a changing technology as the Navy of the nineties was built upon changing weapon systems. How then can we find the means to accept with less pain to ourselves and less damage to our social organization the dislocations in our society that are produced by innovation? I cannot, of course, give any satisfying answer to these difficult questions. But in thinking about the case study before us, an idea occurred to me that at least might warrant further investigation by men far more qualified than I.

A primary source of conflict and tension in our case study appears to lie in this great word I have used so often in the summary, the word "identification." It cannot have escaped notice that some men identified themselves with their creations — sights, gun, gear, and so forth — and thus obtained a presumed satisfaction from the thing itself, a satisfaction that prevented them from thinking too closely on either the use or the defects of the thing; that others identified themselves with a settled way of life they had inherited or accepted with minor modification and thus found their satisfaction in attempting to maintain that way of life unchanged; and that still others identified themselves as rebellious spirits, men of the insurgent cast of mind, and thus obtained a satisfaction from the act of revolt itself.

This purely personal identification with a concept, a convention, or an attitude would appear to be a powerful barrier in the way of easily acceptable change. Here is an interesting primitive example. In the years from 1864 to 1871 ten steel companies in this country began making steel by the new Bessemer process. All but one of them at the outset imported from Great Britain English workmen familiar with the process. One, the Cambria Company, did not. In the first few years those companies with British labor established an initial superiority. But by the end of the seventies, Cambria had obtained a commanding lead over all competitors. The President of Cambria, R. W. Hunt, in seeking a cause for his company's success, assigned it almost exclusively to the labor policy. "We started the converter plant without a single man who had ever seen even the outside of a Bessemer plant. We thus had willing pupils with no prejudices and no reminiscences of what they had done in the old country." The Bessemer process, like any new technique, had been constantly improved and refined in this period from 1864 to 1871. The British laborers of Cambria's competitors, secure in the performance of their own original techniques, resisted and resented all change. The Pennsylvania farm boys, untrammeled by the rituals and traditions of their craft, happily and rapidly adapted themselves to the constantly changing process. They ended by creating an unassailable competitive position for their company.

How then can we modify the dangerous effects of this

word "identification"? And how much can we tamper with this identifying process? Our security — much of it, after all — comes from giving our allegiance to something greater than ourselves. These are difficult questions to which only the most tentative and provisional answers may here be proposed for consideration.

If one looks closely at this little case history, one discovers that the men involved were the victims of *severely limited* identifications. They were presumably all part of a society dedicated to the process of national defense, yet they persisted in aligning themselves with separate parts of that process — with the existing instruments of defense, with the existing customs of the society, or with the act of rebellion against the customs of the society. Of them all the insurgents had the best of it. They could, and did, say that the process of defense was improved by a gun that shot straighter and faster, and since they wanted such guns, they were unique among their fellows, patriots who sought only the larger object of improved defense. But this beguiling statement, even when coupled with the recognition that these men were right and extremely valuable and deserving of respect and admiration — this statement cannot conceal the fact that they were interested too in scalps and blood, so interested that they made their case a militant one and thus created an atmosphere in which self-respecting men could not capitulate without appearing either weak or wrong or both. So these limited identifications brought men into conflict with each other, and the conflict prevented them from arriving at a common acceptance of a change that presumably, as men interested in our total national defense, they would all find desirable.

It appears, therefore, if I am correct in my assessment, that we might spend some time and thought on the possibility of enlarging the sphere of our identifications from the part to the whole. For example, those Pennsylvania farm boys at the Cambria Steel Company were, apparently, much more interested in the manufacture of steel than in the preservation of any particular way of making steel. So I would suggest that in studying innovation, we look further into this possibility: the possibility that any group that exists for any purpose — the family, the factory, the educational institution — might begin by defining for itself its grand object and see to it that that grand object is communicated to every member of the group. Thus defined and communicated, it might serve as a unifying agent against the disruptive local allegiances of the inevitable smaller elements that compose any group. It may also serve as a means to increase the acceptability of any change that would assist in the more efficient achievement of the grand object.

There appears also a second possible way to combat the untoward influence of limited identifications. We are, I may repeat, a society based on technology in a time of prodigious technological advance, and a civilization committed irrevocably to the theory of evolution. These things mean that we believe in change; they suggest that if we are to survive in good health we must, in the phrase that I have used before, become an "adaptive society." By the word "adaptive" is meant the ability to extract the fullest possible returns from the opportunities at hand: the ability of Sir Percy Scott to select judiciously from the ideas and material presented both by the past and present and to throw them into a new combination. "Adaptive," as here used, also means the kind of resilience that will enable us to accept fully and easily the best promises of changing circumstances without losing our sense of continuity or our essential integrity.

We are not yet emotionally an adaptive society, though we try systematically to develop forces that tend to make us one. We encourage the search for new inventions; we keep the mind stimulated, bright, and free to seek out fresh means of transport, communication, and energy; yet we remain, in part, appalled by the consequences of our ingenuity, and, too frequently, try to find security through the shoring up of ancient and irrelevant conventions, the extension of purely physical safeguards, or the delivery of decisions we ourselves should make into the keeping of superior authority like the state. These solutions are not necessarily unnatural or wrong, but they historically have not been enough, and I suspect they never will be enough to give us the serenity and competence we seek.

If the preceding statements are correct, they suggest that we might give some attention to the construction of a new view of ourselves as a society which in time of great change identified with and obtained security and satisfaction from the wise and creative accommodation to change itself. Such a view rests, I think, upon a relatively greater reverence for the mere *process* of living in a society than we possess today, and a relatively smaller respect for and attachment to any special *product* of a society, a product either as finite as a bathroom fixture or as conceptual as a fixed and final definition of our Constitution or our democracy.

Historically such an identification with *process* as opposed to *product,* with adventurous selection and adaptation as opposed to simple retention and possessiveness, has been difficult to achieve collectively. The Roman of the early republic, the Italian of the late fifteenth and early sixteenth century, or the Englishman of Elizabeth's time appears to have been most successful in seizing the new opportunities while conserving as much of the heritage of the past as he found relevant and useful to his purpose.

We seem to have fallen on times similar to theirs, when many of the existing forms and schemes have lost meaning in the face of dramatically altering circumstances. Like them we may find at least part of our salvation in identifying ourselves with the adaptive process and thus share with them some of the joy, exuberance, satisfaction, and security with which they went out to meet their changing times.

I am painfully aware that in setting up my historical situation for examination I have, in a sense, artificially contrived it. I have been forced to cut away much, if not all, of the connecting tissue of historical evidence and to present you only with the bare bones and even with only a few of the bones. Thus, I am also aware, the episode has lost much of the subtlety, vitality, and attractive uncertainty of the real situation. There has, too, in the process, been inevitable distortion, but I hope the essential if exaggerated truth remains. I am also aware that I have erected elaborate hypotheses on the slender evidence provided by the single episode. My defense here is only that I have hoped to suggest possible approaches and methods of study and also possible fruitful areas of investigation in a subject that seems to me of critical importance in the life and welfare of our changing society.

3

Data Processing in a Bureau Drawer

My original intent was to provide you at the beginning of your deliberations with a definition of the word "bureaucracy" that you could have in mind throughout your next months of reading and discussion and to give besides some description of the dynamics of bureaucratic action.* To this end I did what men in my profession do, I looked at bibliographies and professional journals and books and consulted my colleagues. What I wanted was a definition of bureaucracy.

These are some of the definitions that my researches turned up. Bureaucracy is a system; it is a set of rules; it is status and paper work; it is a little world by itself; it is any office with a rug, a separate budget, two dictating machines, eight telephones, ten file cabinets, and three more secretaries than the organization chart provides. None

* This essay was originally prepared in 1957 for a symposium of students at the College of William and Mary. It was to serve as an introduction to their study, throughout the term, of the nature of bureaucracy.

of these, with the exception of the last, is, as you can see, very precise, and none of them is at all serviceable as a basis for our discussion. Toward the end of my researches I then found this statement by a student of the British Civil Service: "Virtually any comment," he said, "one cares to make about this bureaucracy can be shown to be founded on fact." When I found this I gave up consulting scholarly works and turned to the dictionary.

There I found that bureaucracy meant government by bureaus, which is partly true but not very helpful. But I also found that the root of the word is the Latin word *burra. Burra* means sheep shearings. Since sheep shearings are woven into cloth, and cloth in the old days was put on top of desks, and since desks were used in the administration of empires, desks became bureaucracy. I should note here too perhaps that *burra* is the root of another word sometimes associated, especially in the works of Gilbert and Sullivan, with bureaucratic action. It is but a series of easy steps from *burra* as sheep shearing to wool gathering, to wooly-headed, to nonsense, to burlesque.

One thing I did get out of all these investigations. I asked perhaps ten friends of mine for a good working definition of bureaucracy. In these answers I noticed some interesting things. First, they all knew what I meant when I said the word, but not one gave me a definition. Instead, they all responded by giving me an anecdote about some bureaucratic situation in which they had been involved. These anecdotes were usually amusing, always highly personal, and frequently sardonic in tone. They were similar in spirit and content to the description given me some years ago by a Navy captain. He had tried by wit, guile, maneuver, and desperate means to get somebody to read a memorandum he had written in which he proposed some change in departmental procedure. Recounting his failure, he wistfully concluded, "Reforming the Navy Department is like kicking around a forty-foot sponge."

This brings me to the first point I would like to make. I began to realize as I pursued my researches that in talking about bureaucracy one is talking about something as difficult to bring within defined limits as a forty-foot sponge. The instinct of my friends to whom I put my request for definition was entirely sound. In avoiding direct answer and falling back on personal anecdote they were revealing a very important point about the nature of bureaucracy. It is only in part what the dictionary says — government by bureaus — only in part an administrative system and organizational structure. That is why it has proved, I think, so difficult for me to lay my hands on our subject. You can put a nice set of boxes connected by lines on paper; you can write a description of cognizances and duties assigned; you can produce a flow chart of bureaucratic information; but when you are finished you have only part — and I suspect the least part — of the nature of bureaucracy. For part of bureaucracy is an environment, an atmosphere, a pervasive state of mind. This I think is why my friends replied to my question with personal anecdotes, and I suspect also it is why the best descriptions of the nature of bureaucracy are to be found not in scholarly studies but in novels. Novelists can deal with particular situations and the galaxy of human emotions that group themselves within a situation better than students of history, government, or political science. They can make one feel as well as think about what happens in the bureaucratic atmosphere. For this reason I hope you will try to read in the course of this symposium a book called *Delilah* by Marcus Goodrich, or one of the parliamentary novels of Anthony Trollope, or above all, the fullest, best, and most terrifying study of bureaucracy I know — *The Trial* by Kafka.

Having put the definition of the subject this way, that bureaucracy is as much a state of mind or atmosphere as an administrative system, the question remains of how, in the time at our disposal, can we best make an entry into the atmosphere?

I think the best way is to hit the forty-foot sponge a few glancing blows, to take up and examine two or three things that come to anyone's mind when he is presented, as in a blot test, with the word bureaucracy. By looking at these particular things, with a little supporting historical evidence thrown in, it may be possible to reach a tentative conclusion or two at the end. I should say that most of the historical evidence will be drawn from the records of the Army or Navy. This is only because I am more familiar with the history of the Armed Forces, but I think there are advantages in using this kind of evidence anyway. Military situations are more definite, clear, unqualified than those in civil life. The military world in peace is a little like the real world put under laboratory conditions. The effects of experiment can be better studied under these conditions. Such distortions as may occur will not, I think, damage the essential points I want to make about the operations of any bureaucracy, whether it regulates an armed force, a society, a corporation, or a university.

The first thing that comes to anyone's mind when he hears the word bureaucracy is, I suspect, paper. As we have already seen, the word itself derives from a Latin word that has to do with writing at a desk. Much of the folklore of bureaucracy is built up around paper. You all know the old joke — take these papers, file them alphabetically, and burn them. It is supposed to suggest the essence of the bureaucratic method. There is an even better joke that is true and tells much more. Sixty years ago the most powerful officer in the United States Army was a man trained in the practice of medicine who had never commanded a troop in the field in war or peace. He had been made a major general because he had invented a card index. The circumstances were these. In the first Administration of Grover Cleveland, the officer, then a lieutenant, had been assigned to the Bureau of Pensions. At the time of his arrival congressmen were sending over, almost every day, requests for information about the war service of constituents who were seeking pensions. It took, ordinarily, six months for the Bureau of Pensions to answer these congressional requests. One year after Lieutenant Ainsworth's appearance the desired information was in the hands of congressmen within twenty-four hours after the requests had been made. The instrument of this revolution was the card index Ainsworth had devised.

Having thus established his reputation and incidentally his good relations with congressmen, Ainsworth went from strength to strength. By 1905 he was the adjutant general in the War Department in control of the office through which all the paper — records and correspondence — of the whole department passed.

This was a very important position, and if you understand its nature, you will, I think, understand much about our subject. The adjutant general served as a clearinghouse for Army paper in this way. Into his office came, in a given week let us say, a letter from the Corps of Engineers about the construction of a barracks at Fort Russell in Wyoming; into it came a letter from the Signal Corps about a telegraph system to be installed in the barracks. Into it came orders from Personnel detaching Officer Brown from Fort Russell, and into it came papers prepared in the Supply Corps for the travel remuneration of Officer Brown. Perhaps fifty orders and papers bearing on the welfare of Fort Russell entered the office of the adjutant general in a week. That Fort Russell operated smoothly, that only one barracks, not three, was built in the precise spot selected by Army surveyors, that the telegraph was installed the moment the barracks was finished, that Officer Brown was detached the moment Officer Jones arrived to relieve him, and that Officer Brown left with proper pay and travel vouchers — all these were in large part the result of the skill of General Ainsworth's office, which organized all these loose pieces of information into a nicely fitted pattern of activity for the energies at Fort Russell.

As I have said, the career of General Ainsworth reveals a good deal about bureaucracy. His reputation was made because he was skillful in the collection, the filing, and the organization of paper. But there is more to say than that. Notice that the paper had come from a great many different places — Personnel, Engineers, the Signal and Supply Corps. Reflect also that on the paper was written different kinds of information about men and money and machinery and building material. Notice finally that all this information was fitted into an intelligible pattern within which a whole community could act sensibly and in harmony together. This is what a bureau is designed to do. This is what made General Ainsworth a great bureaucrat. He recognized he had to acquire information, to file it in such a way that it could be got at easily, and that it then had to be put together into a scheme of related parts so that others could take decision and action within the scheme. Here, it seems to me, is discovered the real point of bureaucracy — it is a data-processing machine, like the celebrated Univac. It is a computer system. Like Univac, it is designed to take in and digest different pieces of information in far greater quantity than the human head alone can hold and produce from this mass of differentiated data, a useful synthesis.

Let me give another example. When I was in college twenty-five years ago, I was frequently told that the president ran the whole place with the aid of a secretary and a little black book in which he kept the names of the faculty and the subjects they taught and their salaries. That was all the information he needed to make sensible educational policies. Today that college has not only a president but a provost and a vice president and deans, special assistants, and directors, and *ad hoc* committees and standing committees without number. It has, in other words, a bureaucracy. People sometimes wonder today wistfully why the president can't get rid of the deans and buy a little black book. He can't because the college is far larger, because on the campus there are research centers supported by foundations, and laboratories in which the government has an interest, because new departments and intellectual fields not even dreamed of in President Lowell's day have been created. What I am saying is that this college world, like our own, has grown larger, more complicated, more diversified. No single head and no black book can hold the information that is necessary to make intelligent decisions of policy for the whole community. That is why the bureaucracy develops in any social system — to collect, retain, and supply this information in an orderly way.

There are two other observations worth making here about paper. The information I have been talking about is written on paper. This has advantages. It means you can retain it; you don't have to rely on your memory. It means you can file it so you can get right at it when you want it. It means you can move it around and fit it together in different ways for different purposes. Because you can control paper, in other words, you can control the information you need to make decisions. But there are some disadvantages. First, since so much time is spent by people in bureaus working with paper, they may come to set too much store by it. They may become absorbed in receiving it, initialing it, routing it, filing it, keeping it; they may forget to read it in this process. Paper may in their eyes become more important than what is written on it. This is a natural tendency — paper is durable, tangible, easy to manipulate. It is something to see, feel, touch. Information and ideas are volatile, hard to handle, invisible, and they may not even be used. Men in bureaus are not different from men anywhere; they would rather risk their lives and reputations in keeping track of something solid and inert than of something impalpable and invisible. So they may tend to worry more over where a paper is than what has become of the things written on it.

There is something else about bureaucratic paper worth noticing. There are some things you cannot write on it, things any sensible man has to take into account. You can, for instance, write out orders for Lieutenant Brown to leave Fort Russell and report to Fort Ethan Allen, but you can't get on the paper how the lieutenant may feel about it. All kinds of qualifying, modifying, distorting considerations have to be left out of the information written on bureaucratic paper. It is difficult to introduce a sense of urgency, of uncertainty, of change, of growth, of all those strange feelings and attitudes that enter into and disturb any human situation. Concern for paper, in other words, may tend to drive out concern for the human being. Certainly it tends to place a barrier between the bureau and the situation with which it is dealing. It tends to wall off the sense of immediacy and action. No doubt this is one of the reasons why so many bureaucratic papers are written in the third-person singular neuter and in the passive voice. "In compliance with communication 124/35 the following statement has been prepared and submitted although it is recognized that under existing conditions as described in communication 123/08 . . ." Such partial sentences create often enough the impression that no one wrote them at all, that they just appeared.

Perhaps I should add one more thing about paper before I leave the subject. Since it can be kept, the information on it can also be kept. The tendency may develop to value it and use it just because it is there. This may lead to the tendency to protect the hoard of information you have accumulated by the simple device of continuing the policy based on this information, even though the need for the policy is now passed.

And this leads me to the second thing that occurs to the mind when it is exposed to the word bureaucracy and that is "red tape," or what in more dignified terms is called "regulations." Regulations are, in one way, simply the generalizations drawn by the bureaus from the information they acquire. Taken together, a set of regulations provides the pattern of behavior for the energies bureaus are set up to regulate. For instance, the Securities Exchange Commission gets information from brokers, bankers, federal officials, economists, and businessmen. From this information it produces a set of rules to regulate the behavior of those engaged in market operations. Regulations are a way of keeping a system of energies working in harmony and balance. They are an indispensable means for ordering as complicated and delicate and large a society as ours. What else can one say about them?

First, it is easier to make a regulation than to abolish it. Sixty years ago there was in the United States Army a thing called the Muster Roll. It contained the service biography of each soldier. By regulation each month each Army post sent into the War Department additional material to be added to the biographical record of each soldier on the Muster Roll. In 1913 someone proposed to consolidate the information on this roll with other biographical records that were being kept in the medical and pay offices. Money and time would be saved, and no loss of information would occur. Now the Muster Roll, like its name, had a hallowed sound and tradition. Some men had spent their whole service lives in assembling and entering the data on the roll; others at far-flung Army posts found happy occupations in the idle days of peace in preparing the monthly reports. Routines, reputations, methodologies, administrative agencies, and even lives had been built up around the maintenance of this regulation. A fight over the proposal to abolish it took place — so intense that one general was forced to resign from the service before a change could be made.

Second, it is easier to conform to a regulation, even when it is inappropriate to do so, than it is to seek a sensible exception. There is an entertaining case in point that comes down to us from the British Navy of the nineteenth century. There was a regulation in those days that all ships at sea should conduct a firing practice once a year. No one on a ship at sea really liked this regulation. Target practice took up a whole day of unaccustomed hard work; it did little or nothing to improve the marksmanship of jolly tars; it also dirtied the guns and fouled the bright work. So the tendency was to put the whole thing off as long as possible. One ship on independent duty put it off until the last day, when it was discovered to the consternation of the captain and the gunnery officer that water, leaking through the hatches of the ammunition storeroom, had sadly deteriorated the powder supply. What to do? A target practice required, a report on the practice due at the Admiralty, and no practice possible. Simple enough perhaps to explain the circumstances, but ingenuity sought another way. The deteriorated powder was first put overside. Then the gunnery officer took the Admiralty forms, and, with what must have been an interesting set of calculations, filled in scores for the guns of various caliber.

So far, perhaps, so good, though with some slight straining of the conduct becoming to an officer. But then the officer did a really clever thing. He tore the report into small bits and put the bits into an Admiralty dispatch box, and then he added to the box six cockroaches. The whole was sent on to the Admiralty.

In due time, perhaps a year, in the course of which reports of the target practices of all Her Majesty's ships were borne home from foreign waters, filed by diligent clerks at the Admiralty, certainly counted, and possibly studied by officers on duty in the Admiralty — at the end of a year of such orderly activity a dispatch was received from London by the ship that had sent the cockroaches in. It reported that the ship's record for target practice in the previous year had been destroyed in transit by cockroaches and requested a copy of the ship's copy for the record. Back to London, in time, went word that the gunnery officer who had prepared the report in question had been transferred to another ship, taking, apparently, the copy with him. There, so far as history relates, the matter was left.

There are many interesting things that could be extracted from this small anecdote. Here, it is enough to notice again that within the bureaucratic structure the mood of conforming to a regulation rather than seeking a sensible exception becomes highly developed. In passing, it can be said that this may suggest the difference between administration by a bureau and administration by an executive. The executive knows, if he is wise, that he exists to make sensible exceptions to general rules. The bureau knows it exists to make rules. And the tendency to conform to the rules, to regulations, as in this case, leads often to the maintenance of the appearance of things when things are, in fact, quite different. This can, as again in this case, be funny, but in other cases it can be dangerous. Exceptions are made in the interests of making a particular situation conform to reality when the general rule does not cover what is real. So in such instances if one conforms to the regulation, one tends to get out of touch with reality.

Carried to an extreme, this faith in regulation transforms the bureau into a world of its own in which forms, procedures, rules, and regulations have the force of a kind of contrived natural law. Men live in it as though it were real, as though there were no gunnery report not because there was no gunnery, but because there were cockroaches. All this is nicely said by Tolstoy in describing how things went forward in the Russian bureaucracy, a bureaucracy so far detached from the true condition of things it may be said to have been in a catatonic state for a whole century. The passage is a little long, but it has so much to say to our purpose:

It happened that the famous Commission of the 2nd of June had set on foot an inquiry into the irrigation of lands in the Zaraisky province, which fell under Alexey Alexandrovitch's department, and was a glaring example of fruitless expenditure and paper reforms. Alexey Alexandrovitch was aware of the truth of this. The irrigation of these lands in the Zaraisky province had been initiated by the predecessor of Alexey Alexandrovitch's predecessor. And vast sums of money had actually been spent and were still being spent on this business, and utterly unproductively, and the whole business could obviously lead to nothing whatever. Alexey Alexandrovitch had perceived this at once on entering office, and would have liked to lay hands on the Board of Irrigation. But at first, when he did not yet feel secure in his position, he knew it would affect too many interests, and would be injudicious. Later on he had been engrossed in other questions, and had simply forgotten the Board of Irrigation. It went of itself, like all such boards, by the mere force of inertia. (Many people gained their livelihood by the Board of Irrigation, especially one highly conscientious and musical family: all the daughters played on stringed instruments, and Alexey Alexandrovitch knew the family and had stood godfather to one of the elder daughters.) The raising of this question by a hostile department was in Alexey Alexandrovitch's opinion a dishonorable proceeding, seeing that in every department there were things similar and worse, which no one inquired into, for well-known reasons of official etiquette. However, now that the glove had been thrown down to him, he had boldly picked it up and demanded the appointment of a special commission to investigate and verify the working of the Board of Irrigation of the lands in the Zaraisky province. But in compensation he gave no quarter to the enemy either. He demanded the appointment of another special commission to inquire into the question of the Native Tribes Organization Committee. The question of the Native Tribes had been brought up incidentally in the Commission of the 2nd of June, and had been pressed forward actively by Alexey Alexandrovitch as one admitting of no delay on account of the deplorable condition of the native tribes. In the commission this question had been a ground of contention between several departments. The department hostile to Alexey Alexandrovitch proved that the condition of the native tribes was exceedingly flourishing, that the proposed reconstruction might be the ruin of their prosperity, and that if there were anything wrong, it arose mainly from the failure on the part of Alexey Alexandrovitch's department to carry out the measures prescribed by law. Now Alexey Alexandrovitch intended to demand: First, that a new commission should be formed which should be empowered to investigate the condition of the native tribes on the spot; secondly, if it should appear that the condition of the native tribes actually was such as it appeared to be from the official documents in the hands of the committee, that another new scientific commission should be appointed to investigate the deplorable condition of the native tribes from the — (1) political, (2) administrative, (3) economic, (4) ethnographical, (5) material, and (6) religious points of view; thirdly, that evidence should be required from the rival department of the measures that had been taken during the last ten years by that department for averting the disastrous conditions in which the native tribes were now placed; and fourthly and finally, that that department be asked to explain why it had, as appeared from the evidence before the committee, from No. 17,015 and 18,308, from December 5, 1863, and June 7, 1864, acted in direct contravention of the intention of the law T . . . Act 18, and the note to Act 36. A flash of eagerness suffused the face of Alexey Alexandrovitch as he rapidly wrote out a synopsis of these ideas for his own benefit. Having filled a sheet of paper, he got up, rang, and sent a note to the chief secretary of his department to look up certain necessary facts for him. Getting up and walking about the room, he glanced again at the portrait, frowned and smiled contemptuously.*

One final thing about regulations, they tend to multiply. The reason is probably obvious, but I will give a brief example. My oldest daughter broke a horse to the harness last summer. She began by teaching him to start at a given signal and to turn right and left at the pull of the reins. She soon found, of course, that it was not enough to start a horse; you had to stop it too. So another procedure was added. Then she went on to train it to back up and turn around and to walk and trot but never to canter. By the time she was finished, the poor animal was caught and held in a whole net of rules and procedure. It is the same with any procedures designed to organize human energy in a pattern of activity. In order to make the pattern work, one seeks to eliminate every uncertainty and variable that might disturb the scheme. So the tendency in every regu-

* Constance Garnett (trans.), Leo Tolstoy, *Anna Karenina,* Vol. I, Part 3, pp. 343–344. New York, 1939.

lating body is to reach out and extend rules over the whole range of human activity. That is why questionnaires get longer and the set of regulations more detailed. That is also why red tape has its unpleasant connotations. It tends to form an increasingly binding set of rules within which a man, an agency, a company has to operate.

Now let us turn to the third and last thing that I think comes to mind when one hears the word bureaucracy. This is harder to put in one word than paper and regulations. Perhaps it is most suggestively contained in the word "they." I thought of this when one of my friends, hearing that I was coming down here to talk on this subject, said, "Tell them who 'they' are." One of the things about "they" is that they are hard to find and identify. I knew a man who some years ago tried to find out who had charge of the curriculum at one of our great service academies. He went to every section in the department that might conceivably have something to do with education. In these places he found no one who knew about the curriculum or could put him in touch with someone who did. Finally in an odd corner he came upon a civilian, an aged little man who had been for years in the department and who years before had been assigned the task of incorporating a change or two that had been decided upon in the curriculum. Since then he had, in his anonymous position, taken charge of the curriculum.

Another example taken from a large manufacturing concern comes to mind. The company was, like others, divided into separate parts — sales, finance, design, and so on. Regulations for the coordination of these parts were set down in a company procedures manual. Those who controlled this manual obviously played an important part in the operations of the company. The men who had this important place were members of a small section of the

special duties, its nice categories, its ordered set of rules, its rituals, its familiar faces, its known and definable limits — with all these a bureau is indeed like a comfortable and sheltered world. But even in a comfortable and sheltered world men have desires and ambitions and prides. Within the confines of the bureau men sometimes lose themselves in a struggle for position and power, and concentrate on ways to improve their own position in it.

Let us go back for a moment to the Muster Roll to illustrate this point. The struggle over the Muster Roll began in this way. The President of the United States had appointed a Commission on Economy and Efficiency in the Government. The Commission had examined every part of the government and made recommendations to improve the conduct of the government's business. In the War Department it had discovered an undue preoccupation with paper work. In recommending ways to improve this condition, it suggested the consolidation of the Muster Roll with other records. The adjutant general in whose office the Roll was prepared fought the recommendation with every means at his disposal. First, when the Chief of Staff sent him a memorandum about the Muster Roll, the adjutant general did nothing, a method that frequently works in large agencies. In due course, the Chief of Staff sent another memorandum. To this the adjutant general sent a memorandum defending, with elaborate argument and statistics, the Muster Roll as an indispensable instrument of government. Then he went on to say some very rude things about the intelligence of the Chief of Staff since it was obvious he could not understand that the Muster Roll was indispensable. The Chief of Staff then took this memorandum to the Secretary of War, who was so angry at the adjutant general for using rude language that he prepared a court-martial. Before the court-martial could meet, the adjutant general resigned, and the Muster Roll was thus eliminated.

Notice these points about the episode. First, the struggle was ostensibly about the Muster Roll. Actually, it was a struggle over the question of who would have the controlling power in the War Department — the adjutant general, who controlled, as we have seen, the paper in the department, or the Chief of Staff, who was charged with planning for war and maintaining military efficiency. This was an old, old struggle. For years adjutants general and Chiefs of Staff had been fighting each other. They had fought over whose office should be next the Secretary's; they had fought over who ran recruiting stations; and they had fought over which one could assign officers to duty. For instance, once a Chief of Staff had ordered an officer from Washington to Oregon on special duty, and the adjutant general had refused to authorize payment for travel. The Muster Roll was an incident in this struggle which over ten years caused one Chief of Staff to resign and another to go to a hospital. For ten years this contest for authority within the bureau had paralyzed large sections of the War Department.

Now the interesting thing is that nobody in the course of this contest admitted what it was about. It was fought over travel money and assignments to duty and Muster Rolls, not as an admitted contest for power. And it was decided, you may have noticed, not even on the merits of the case, the usefulness of the Muster Roll. It was decided by the Secretary of War on the grounds that the adjutant general had used rude language. All this had its entertaining qualities as has the world of Alice in Wonderland, but it should be pointed out that while the conflict of wills was going on in this strange world, the real world of the Army was left to suffer. Nobody had much time to think about its particular needs, whether damage was done to it because the officer had not gone from Washington to Oregon or whether it really needed a Muster Roll.

One more point about the Muster Roll before we reach some concluding remarks. Notice that the issue was raised by an outside agency, the President's Commission, and solved by an outside agency, the Secretary. Recall that in the earlier case of the procedures manual in the company, the solution was provided by a new president and a reorganization. This seems to me important. The bureau as a self-enclosed world has great trouble in changing itself. It tends to settle into a firm structure within which rival passions and energies can contend with each other. It is highly dependent upon outside stimuli to force changes upon itself. Everyone inside is too committed to the special world.

Now what does this disjointed set of anecdotes and reflections add up to? I would hope at least that you have received some general impressions. The first impression would be this: that bureaucracy is an attempt to give useful expression to some part of man's own spirit. It is an expression of man's effort to act as a rational being, to fashion for himself out of observed data sensible conclusions, the plans and programs and hypotheses on which he can make decisions and take wise action. It is an expression of man the political or social animal, his desire to work and cooperate with others within a general scheme and his desire to erect general rules that will protect the community from the randomness of individual energy. It is an expression of his intense need to introduce order, consistency, and predictability into the conduct of his affairs. It is an expression of that part of man which seeks to reduce the danger that surrounds any human life, that part of him that wants to play it safe.

I hope also that you have received another impression,

that bureaucracy, insufficiently controlled, tends to exclude another part of man's own spirit. Though man is rational, he is also biological and emotional. He proceeds not only by his wits but by his instinct, his intuition, and his feelings. And though he is political and social, he is also a single person and idiosyncratic. He has peculiar personal needs and fears and desires, and retains a sense of himself alone. Though he has the urge for order and safety, he also persists in inconsistency and the disorderly and even dangerous conduct out of which come so many of the truly imaginative and original works of man. Within the bureaucratic system is a spirit antagonistic to this part of man. It is the awareness of this, I think, that produces the sardonic or bitter or uncomfortable tones that so often appear when people speak of bureaucratic action.

I hope too you have received a third impression: that something happens to a plan of action or a program when you enclose it in an institution, or in this case, in a bureau. It tends to lose its freshness, its responsiveness, within the hard and rigid lines of the institution. Those at work on the program tend to get caught in complicated struggles for power and place within the structure and may forget the purpose for which they exist. They find themselves in the position of the Secretary of the Navy in the nineteenth century who is said to have complained that just when he had gotten everything working smoothly in his department, a war came and spoiled it all. Any institution, any department, any bureau that gets thus lost in its own concerns loses its awareness and responsiveness to the outside situations it is supposed to deal with. This is another reason why men speak sardonically or fearfully of bureaucracy. They feel that power over them may be in the hands of men who don't care about them, who may not even be aware of them.

These are the general impressions I hope I may have left with you. Now in conclusion I would like to say a few words about why I think you should think about these matters, why you have this symposium on the subject. There is a rising anxiety, which I think the fact of this symposium suggests, that we are today insufficiently protected against the restless urge of the bureaucratic spirit to extend itself over all our affairs. There is an uneasy feeling that in this complicated, distracted world our need for order and safety may drive us into the bureaucratic situation.

Certainly there are evidences of this. In the time that the country's population has doubled, the civil servants on the federal payroll have increased seventy times. In the corporations where most of us live and work the bureaucratic structure of administration is far advanced. About us all there are more forms and regulations than there have ever been before. The circle of independence for each man, as John Stuart Mill called it, is being diminished.

Thus far it seems to me it has been justifiably diminished, that such evidences of the bureaucratic spirit as have manifested themselves are necessary in the present contingency to keep our complicated, diversified, intricate, and fragile society from shaking itself to pieces. I can also recognize, however, that these evidences are a portent, as some believe, that we may find ourselves in the world of 1984 before we really get there.

I am myself more optimistic for several reasons. I will state two briefly in the hope we may talk about them at greater length later. What we call bureaucracy, with its interest in fixed and uniform solutions, thrives best in static environments. It could fasten itself on Russia or China in the last century because they were ponderous, agrarian societies, honeycombed with rigid class distinctions. They were very hard to move unless, as happened, they were totally destroyed by revolution. Here we have in our possession an instrument of constant change, science and technology. Our scientific findings translated quickly into technical applications are producing steady disturbance in the environment. New means of production, transport, and distribution are almost daily thrown into society to modify our habits and the regulations by which we live. In such an environment it is at least harder for the bureaucratic spirit of fixed rules to get a toe-hold. Science and technology are the influence from outside I spoke of earlier. They are like the Presidential Commission on Efficiency or the new president who got rid of the group who wrote the procedures manual, constantly interfering to throw the bureaucratic balance of things out of balance.

Then too we have ourselves. In spite of our huge size, in spite of our specialization, in spite of the immersion of the individual in large classes, large companies, large communities of ranch houses, I think it is still possible to believe in our whole and separate selves. If we keep in sensible touch with our heritage built on individual freedoms; if we learn about ourselves in the tradition of liberal education, as opposed to vocational or technical training, the tradition built up out of the creative intelligences of single men; if we keep alive and responsive to the tremendous possibilities of our changing world; if finally we have the nerve to think the single self is worth perpetuating, we have the means to ward off the fate of the organization and the bureaucratic man. Whether we do or not depends, I think, in large part on how frightened we get, and more, on how much we allow our fears to govern ourselves. And that is something, fortunately, I think we can determine for ourselves.

4

The Pertinence of the Past in Computing the Future

I was told by higher authority to talk about the pertinence of the past, or to put it crudely and in the way I think it was meant, to talk about "what possible use is there in history."* I have every intention of obeying my instructions, but I am going to do it in my own way, which is a way I learned from the Queen in *Through The Looking Glass*. She was telling Alice that it was a great advantage to have a memory that worked both ways.

Alice replied that she was sure her memory didn't work in that fashion, that in fact she could not remember things before they happened.

"It's a poor sort of memory that only works backwards," the Queen thereupon remarked.

And I have decided that, at least for the purposes of this

* Talk delivered at a reunion of the Sloan Management Fellows at M.I.T. in 1959.

paper, it is a poor sort of past that only deals with what has happened. So before I am through, I am going to try to recall some of the future.

But out of deference to my title and to those in higher authority I want to begin with a brief survey of a situation that has taken place. That situation occurred in 1943. In March of that year there were, on a normal day, about 85 German submarines operating in the Atlantic Ocean. On the same kind of normal day the United States had available to fight the submarine in the Atlantic Ocean the following forces (the figures are approximate): 4 to 6 small carriers, 125 destroyers, 30 of the new destroyer escorts, 200 small craft (subchasers, converted yachts, trawlers), and, finally, 500 to 600 planes (scouts, fighters, and bombers). In that month of March the submarines sank in the Atlantic Ocean 567,401 tons of merchant shipping.

The half-million tons of merchant shipping lost in March represented the average monthly loss since the German submarine offensive against this country had begun off Hatteras in January 1942. These continued losses, greater at times than the replacement rate of new construction, placed the whole allied war effort in jeopardy. Not only were the ships that were torpedoed lost; the food, oil, and material being borne to the island base of Great Britain and to our troops abroad were lost with them. It seemed probable if not certain in March 1943 that our losses to the submarine would prolong the war, and it seemed at least possible that these losses would preclude the chance of victory.

Therefore a conference was called in Washington in March 1943 to discuss ways to reduce the attrition caused by the submarine. Out of this conference two distinctly different points of view emerged. These views had to do with the way the forces available, ships and planes, should be used to fight the enemy. On the one hand, it was held that all these elements should be retained in the organization of forces as it then existed — that is, the convoy system. The convoy system had been inherited from the previous war. It grouped ships — fifty, sixty, seventy of them — as I am sure you know, into formations protected by an escort force, a protective screen, of ships and planes. These convoys then sailed along sea-lanes determined by the location of the ports of call.

The great advantage of the convoy system was that it solved the problem of search and attack with great economy of means. The fighting ships did not have to chase all over the farm, in a favorite analogy, looking for the hornets; they did not have to look, in another favorite phrase, all over the limitless ocean for the enemy. Submarines seeking out merchant vessels were forced into contact with the protective screen and further forced into action with our fighting ships concentrated at the point of contact as the rules of tactics prescribe.

The support of the convoy was most ably and clearly stated at the Convoy Conference by Admiral E. J. King, the remarkable man who was Commander in Chief of the United States Navy. He said, in yet another favorite analogy, that there was no point in hunting needles in haystacks. Ships and planes were effective only as parts of the convoy system, to provide surface support and air cover. Anti-submarine warfare, he concluded, must concern itself primarily with the escort of convoy.

Against this view was the other. The convoy system, it was said, was not effective. Submarines too often fought their way through the protective screen to destroy the merchant vessels neatly arranged in convoy formation. The continued sinkings proved as much. A new approach was needed, an approach based upon the following considerations. The ocean was not, in fact, limitless for the submarine. Any U-boat based on the Continent would have to pass through narrow waters on the way to the Atlantic hunting grounds. In the second place, it was unnecessary in the modern conditions to hunt for needles in haystacks. Since all submarines communicated by radio every night with the Admiralty in Germany, the positions at sea could be nicely fixed. Moreover, radar-equipped planes could pick up the target of a surfaced submarine — and submarines in those days had to surface frequently — a good many miles off.

Under such conditions, the convoy system was not, as it seemed to be, a conservation of forces; it was more nearly a dispersion of energy. What was needed was the organization of forces — carrier and land-based planes — to search out and destroy submarines from the air before they got anywhere near the merchant vessels. This position was clearly stated, reasoned out, and factually supported by Edward Bowles, a scientific adviser to the Secretary of War on the problem of undersea warfare. He said that "convoying was at best a most inefficient procedure and that aircraft could be used much more effectively in carrying the attack to the enemy wherever he may be found."

Before describing how this difference of opinion was resolved, it will be useful to say a word or two more about the character of submarine warfare. This warfare was not like those old frigate engagements in which a captain could win an action and gain a reputation by the skilled use of something called his seaman's eye or by his recollection that Lord Nelson had once said that "no captain could go very far wrong if he lays his ship alongside the enemy." Although submarine warfare took place on the trackless seas, it did not in fact present a shapeless situation through which men worked their way by old sayings, rules of thumb, and conventional wisdom. In this warfare things like cruising radii, radar capacity, configuration of land masses, ship characteristics, torpedo ranges, submergence depth, numbers of forces, and all the rest of it imposed a precise form upon the exercise. In other words, submarine warfare was, to a degree, a defined and limited system, permitting close analysis and the nice calculation of forces and probabilities.

It would seem likely therefore that the choice between the convoy system and the doctrine of search and attack could be determined by the workings of pure reason. You could take all your forces with their known numbers and characteristics and allocate them in the interests of greatest efficiency within the rigid framework provided by land masses and traditional shipping lanes, and you could do this simply by using your brains.

Unhappily, in the event, it didn't work out quite that way. For four months the argument between those who supported the convoy system and those who advocated search and attack by plane went forward. Those who wanted the convoy system made certain claims; they said, for instance, that the convoy was the only thing that would work and that it would work as soon as enough ships were added to the fleet to make the protective screens stronger. They made certain assertions; they said, for instance, that submarine warfare took place on the sea and that men trained in seamanship understood that kind of war better than men who were trained to fly in the air. They made certain assumptions; they said, for instance, that to give planes independent missions for search and attack would so dislocate command relations between air and surface forces that the control over operations would disintegrate. On the other side it was said that the convoy system had proved inadequate; that planes had already demonstrated their capacity to search and destroy; that naval officers were captives of the past, were indeed, as one proponent of search and attack said, "an indigestible bunch of society admirals and old crustaceans."

What happened in the months of argument was that the discussion left the high level of sensible analysis and degenerated into assertions, irrelevancies, and personal remarks. The conflict was finally settled after an interminable debate over command relations in which the patience of those supporting air search and attack simply wore out. The convoy system was retained as the primary means of protection for the merchant tonnage.

I have drastically condensed and foreshortened the history of this episode. I have drawn from it, in so doing, much of its drama and fascination; I have probably made it seem too "either, or" a proposition between the two sides. But I hope I have not distorted it unduly, and I hope I have made one thing clear. The problem presented was one filled with measurable, quantifiable data susceptible of analysis and the drawing of reasonable conclusions. The argument did not proceed in an orderly way, however; it did not, ordinarily, take account of the data that had been supplied, and it was not resolved as an exercise in pure reason.

The reason why things turned out as they did can be found, I think, in the fact that there was more at work in the problem than the quantifiable data and the measurable energies. There was first the influence of memory. The convoy system in its modern manifestation was the product of the First World War. In the early days of that war when the submarine was a novel weapon, all kinds of ways to attack it were proposed and tried — the patrol system, the mystery ship, the North Sea Mine Barrage, the bombing of the construction yards at Zeebrugge. Naval officers were constantly being bombarded by politicians, civilians, and writers to the newspapers with bright ideas. Many, as I say, were tried, and all were discovered to be in one way or another expensive and futile dispersions of energy. Nothing really worked well until the convoy system was introduced. Deep in the consciousness of naval officers therefore was burned the recollection that it alone had worked. So when men in 1942 and 1943 came up with bright ideas — some old, like mystery ships, some crazy, like arming seals with warheads, some new and substantial, like searching and attacking with radar-equipped planes — it all seemed like an old familiar and sad story. It mattered not that in thirty years planes had increased their potential and radar had been invented. The conditioned reflexes and unhappy memories of previous experiment interfered with the cool and wholly rational calculations of present possibilities.

There was a second influence at work in the situation. The primary pressure for air search and attack came from the Army. In the first days of the war the Navy with insufficient planes of its own had borrowed some Army aircraft. These planes, land-based bombers, flown by Army aviators and under Army administrative command, but under the operational control of the Navy, had roused the Army's interest in the problem of anti-submarine warfare. Trained to attack specific targets, fascinated by the new bombsights and radar devices, uninhibited by memories of the convoy system, the Army had explored new doctrines. In the spring of 1943 the Army proposed that their own planes should be used on independent missions of search and attack. In so doing they were in fact offering to share a part and perhaps a determining part in the Navy's own war. Thus, to the natural resistance against a proposal from a sister service was added the threat of a loss of control of the function for which the Navy existed — in other words, a loss of sovereignty. As the principal historian of the Navy has said, "Admiral King had no intention of permanently sharing with the Army what he conceived a naval responsibility."

And there was, less easy to prove, probably impossible to prove with supporting data, but to me at least convincing if speculative, a third influence at work. The major instrument in the doctrine of search and attack was the land-based plane. From the very beginning of things the Navy had been built on the surface vessel — sailing ship, battleship, or carrier, still the surface vessel. All procedures, all customs, all habits of mind, all sources of institutional form and individual power — in short, all social and cultural arrangements — had been constructed upon the ship. The proposal to make the plane not a supplementary element to supply air cover for a convoy, but a dominant weapon suggested a change not merely in submarine warfare but a change, perhaps a radical change, in the naval society with all its accumulated meanings and values.

I can make the point of this last part a little clearer perhaps by a brief digression. In a celebrated passage Tocqueville describes what happens to a man in a fluid society where change of residence and status is frequent — that is, in a society like ours. Such a society, says Tocqueville, makes every man forget his ancestors, hides his descendants, and separates his contemporaries from him; it throws him back forever on himself alone and threatens to confine him entirely within the solitude of his own heart. On the other hand, in a society where families remain for centuries in the same condition, often on the same spot, all generations become, as it were, contemporaries. A man knows, or feels he knows, his forefathers and respects them; he thinks he already sees his remote descendants, and he loves them. He willingly imposes duties on himself toward the former and the latter, and he will frequently sacrifice his personal gratifications to those who went before and to those who will come after.

A man in these circumstances as Tocqueville describes them discovers satisfactions and meanings as a temporary custodian of principles and purposes maintained in a great human continuum. He is part of a great network of communications and associations in which he knows others and, even more important, in which he is known by others. This is one of the ways to find out who you are. Satisfactions and meanings of precisely this kind are built into the naval service. A young officer told me not long ago that one of the rewards for living in the Navy was the feeling that you were in a relay race, carrying a torch taken from those who had gone before — some of those indeed were heroes — and passing the torch to those who would come on the Navy list in the future. He was, in other words, in a human continuum where all members, past, present, future, became as it were contemporaries, held together in a network of association and by enduring allegiances to common ideals, as the image of the torch implies, of purpose and service. It is such allegiances and awarenesses that hold an armed force together in the empty times of peace and in the dangerous vicissitudes of war. If the naval society is threatened, as by sharing command of its function with the Army or by the sudden displacement of its familiar means — the surface ship — by a new instrument — the land-based plane — why then all these allegiances would be weakened or even cancelled out. What had been an enduring, self-confident civilization would then, quite possibly, disintegrate into a lonely crowd.

The purpose here is not to decide the merits of the case for or against the convoy system, nor yet to argue for or against the continuation of the Navy as a peculiar institution in the precise form it had assumed in 1942. The purpose of this extended anecdote was only to demonstrate that in any human situation, no matter how filled with quantitative data it may be, there are always present powerful human considerations that are incommensurable. These incommensurables — a tangle of memories, prejudices, emotional needs, aspirations, common decencies — exert a tremendous and probably always a determining influence upon the real, as opposed to the exposed, nature of a situation. Any wise decision in such a situation must take into account not only the data from which logical conclusions about present operating efficiencies can be drawn, but that other information which leads to the nonlogical understanding of what human beings are, need, and want to be. In the anecdote I have told, the decision arrived at by a calculation of cruising speeds and firepowers may have to give way to, would certainly always have to be modified by, a further calculation of the nonquantifiable needs of the naval society. That is why E. M. Forster, the British novelist, said that the only true history is the history of the human affections. All others, even economic history, he said in an unnecessary aside, are false.

All the foregoing has been said in extended preface to my principal interest in this paper, which is to try to find out what may happen to the affections in the time of the computer. What I want to try to do is to raise a few very tentative propositions, an agenda of the kind of thing we will have to think about as we seek to live at peace with the machine. You all know some of the jobs the computer can do today, how it can prepare payrolls with deductions for the individual case, how it can assist in the control of inventories and the shaping of production schedules. It can, even now, do something more.

For the last ten years I have had a Williamson circuit in my radio amplifier. In its utilization of the power available, in its capacity for internal correction, it faithfully represents the elegant solutions to technical problems that occurred to the man who made it, Williamson. It stands as an admirable expression of his own peculiar understanding and intuitive skill, as an expression, in other words, of himself. Today new circuits for conventional amplifiers like mine can be and are designed by the computer. The special aptitudes, the distinctive bundle of intelligence and affections that Williamson brought to his work — the bundle that was, in fact, Williamson — has been displaced.

Computers, as you know, have other competences. They can discover proofs for theorems of logic; they can solve trigonometric identities; they can do formal integration and differentiation. They can already, or have given clear indication that they will be able to in the foreseeable future, do the following things: remember, learn, discern patterns in loose data, make novel combinations of old information, and, most striking of all, introduce surprise into an intellectual situation. There is still a question about the limits of the computer's imagination — its ability to create something like a Newtonian hypothesis or to construct something like Handel's *Water Music.* But Herbert Simon, a cautious student of these matters, has said that "Insofar as we understand what processes are involved in human creativity — and we are beginning to have a very good understanding of them — none of the processes involved in human creativity appear to lie beyond the reach of computers."

The capacities I have described — remembering, learning, discerning patterns, making surprising combinations of data — are most if not all of the capacities men bring to most if not all industrial problems. If I am correct in saying this, then it must seem that sooner or later the machines will be able to take part in the thinking that goes on at every level of an industrial enterprise; that they will be in a position to influence decisions that will affect the ultimate policies and purposes that govern our industrial life. No less certainly than the civilization of an armed force is based on its weapon system is the civilization of our society based upon the instrumentation of the industrial process. All our economic and social arrangements — how we feel about what we do, which is all that culture is — are founded upon the way our industrial energy is organized. How large a part and what kind of part do we want the computer, with its overriding skill in the rational analysis of the measurable data, to take in the decisions that determine the way this energy will be organized? This is worth thinking about.

There are lots of ways to think about it. If you are a junior executive now you can wonder how soon this machine will invade the ranks of middle management, and if and when it will reach the level of the senior executive if and when you actually reach that level. You can also, if you are me, wonder when you will be out of a job. Because of the cheap reproducibility of computer programs, it will, it is said, become cheaper and easier to transfer new knowledge and skill to machines than to men. As a faculty member I'm not sure it will be cheaper, but as a pedagogue long steeped in the educational process, with its immense inefficiencies and wastes caused by the uncertainties and resistances of human beings, I am quite prepared to believe that it will be easier for computers to teach computers.

production department of the manufacturing division of the company. They had reached their eminence in an interesting way. Long before, when the manufacturing division had been reorganized, this little section had prepared a new set of procedures for the division. By this act they had established themselves as men who understood procedures. So whenever a new regulation was required, the matter, by the pull of bureaucratic gravity, was brought to them. By slow and imperceptible degrees they established a monopoly in procedure. As one of the men in the firm remarked, "The result in time was that a group of uninformed people in an obscure corner of one division were preparing ground rules for all departments and hence controlling very important aspects of our operations. The results," he went on, "were often ridiculous." Once the company had to make a proposal — a bid — on a new piece of business. "The cross-coordination, approval, revision and records activities involved in this procedure were so complex," it was reported, "that the system was almost completely regenerative, and there was actually no way of getting a proposal out of the company." One of the interesting things about this example is that men at work in different parts of the company were rarely aware that they were being frustrated in the performance of their tasks by the existence of the "they" who had from their obscure corner prescribed the over-all coordinating procedure. Not, in fact, until the company was reorganized by a new president was the source of trouble fully understood and removed.

There is another point to be made about "they," about those who work in the bureaucratic situation. It has to do with the fact that a bureau, whether a great public agency, a department in a university, or a large corporation, tends to become a world for those inside it. With its

But matters like this — the temporary technological unemployment of the individual, like you, like me, like Williamson — are inconsequential when laid against what I would think was the primary source of concern about the machine. The computer is no better than its program: the quality of its decision is determined by what is put into it, and men will decide what is put into it. This is a source of hope and delight in one way. It means that like the steam engine, the steel mill, the dynamo, the computer is an opportunity to be exploited, an immensely powerful extension of man's ingenuity and power in the service of his will. But it is also a source, as I have said, of concern. If we put the wrong things into it, if we select the wrong problems or state the right problems incorrectly, we will get unsatisfactory solutions. Perhaps the easiest way to put it is that in using the computer, man will get the answers he deserves to get.

Here it seems to me is the cause for real alarm because of a sad historical fact. On the record men have been luckier in giving answers than in asking the right questions, or at least they have been more skillful in making up half answers they can live with than in putting the full questions accurately framed.

This is not said in the hope that it will sound like an epigram. Let me give some brief examples out of our own short history. In the Constitution of the United States as it was ratified it is indicated that for certain purposes a colored man counts as three fifths of a white man. In the seventies and eighties of the last century bimetallism was accepted as a solution to our monetary difficulties. Yesterday, today, and apparently forever there is the legislation dealing with our farm problem.

These jerry-built solutions were arrived at in part because we did not care, for reasons of policy, expediency, ignorance, or pure fright, to state the problems fully and accurately. Sometimes we did not assess the quantifiable data, as the weight of the silver bloc, correctly; sometimes we were not fully conscious of the load of memory, prejudice, hope, faith, and so forth that had been dumped into the situation; and sometimes, indeed almost always, we did not make the effort to separate out the useful from the useless memory, the good from the evil prejudice, the legitimate faith from the illusion, in the problems presented.

In dealing with the computer, with its incisive capacity to reach clear decisions from the data presented, we will find ourselves in grave difficulties, I think, if we persist in such sloppy definitions of the problems we wish to solve. We will have, in the near future, to ask ourselves if we can assess even quantifiable data without distortion; we will have to painfully consider whether we are ready to lift out our raw affections for observation and analysis, whether we are honest enough and brave enough to make the necessary separations between the indispensable and the irrelevant feeling.

There is an alternative in all this that we ought at least to think about. We can begin from the proposition that man is because he thinks and only because he thinks. That is, we can disregard the existence of the affections and program only the quantifiable, measurable elements that can be dealt with by pure reason. Take for instance the anti-submarine problem in 1943. Feed in the clean elements: radar capacities, firepowers, cruising ranges, and so forth. Just the thing for a computer. Let us assume, as I think might well have been the case, that the clean answer came back: use the land-based Army planes for search and attack in the narrow waters of the Bay of Biscay. Let us then assume further that this sharing of control over operations and the introduction of the plane as a dominant weapon would undermine the naval society, altering not only its form but the structure of value and affection that held it together. This would mean that you had swapped cultural stability, determined in large part by the affections, for immediate operational success as prescribed by pure reason. When translated to the larger sphere of the industrial process and the cultural and social arrangements based upon it, the implications, I should think, are clear.

This is certainly one way to do it, and it may be the way, or at least the way we will take. The history of man can be understood not only as the history of the affections but as the effort of man to use his distinctive instrument, the intelligence, to triumph over the affections, the effort to introduce logic and order into the control of affairs made messy by the affections. Then too, as is well known to all foreign observers, we are a pragmatic people living always for the operational success in the existing moment, unexcited by the past, unaware of any future beyond the next tomorrow, which is another day.

With such ideas in mind, and especially in an era of dynamic technological change, it may be argued that all the accumulated baggage of value and culture created by the affections is irrelevant or unnecessary. The thing to do is to reach logical conclusions in the immediate circumstance and bring such memories and affections as do exist into accommodating new alignments to fit the logical conclusions.

I want to take this proposition one step further. Sigmund Freud had a theory that art is merely a neurotic compromise, an enormous effort by men to give reality in the imagined abstract to desires they could not fulfill in action. It is possible that our affectionate attachment to particular schemes of value are equally symptoms of our incapacity to run an environment in a reasonable way. For instance, in the thirteenth century a man petitioned Almighty God only perhaps because he thought he could control nature in no other way. Or a man in the nineteenth century made hard work, thrift, giving to the poor, and the sacredness of individual enterprise into a scheme of value because they were the only devices that enabled him to deal with an uncertain economy he could not run or understand by his reason alone.

If this view of things is accepted — and it has real pulling power for anybody attracted by economies of means, elegant solutions, and nice analyses — then man can learn to live with the computer that feeds only on the quantifiable, measurable data and that yields up the correct operational decision for the moment. Detaching himself from familiar social, institutional, and personal commitments, he will be able to bring whatever affections he has into a continuous alignment with the ever changing decisions about what will work at this moment in the industrial process.

I think that in the future we should look at this proposition with care, but I must say I don't like it much. It seems to rip man away from the qualities and contexts which in the past we have ordinarily presumed gave point and meaning to his life. It seems to turn him into an accommodating mechanism designed to force whatever fitful feelings he may have into grooves cut to a temporary pattern by the intellectual necessities of a situation. It sounds as though in the interests of survival we would be willing to survive as freemartins, dead to rapture and despair. Maybe I don't like this prospect simply because, as the anthropologists say, I am culture-bound.

But I don't wish to be misunderstood here. I believe, of

course, that the intelligence is one of the determining things about man. I am with Whitehead where he said that today the rule is absolute, the society that does not value the trained intelligence will die. I also believe the computers are here to stay and it's no good trying to melt them down, as they tried to bust up the first power looms or to destroy the first Bessemer converter built in this country. I further believe that the computers supply an opportunity to organize and enrich our society that is greater than the opportunity offered by any other agency since the discovery of fire, which also has the power to destroy. And finally, in this confession of faith, I believe that man is a creature distinguished not only by the intelligence but by the affections as well, which means, I guess, that he is a creature of rapture and despair. But which means also that the affections have an existence, an identity, a set of needs and claims, a shaping influence in the life of man that is their independent own. Man is, not only because he thinks but because he feels, and it is the interaction between these two impressive energies that establishes what people today love to call the human condition. This at least is one of the things I think I have learned from history. So I would add that in confronting the computer, we must examine with care whether the rule is not equally absolute: the society that does not value the educated heart — or wherever the seat of the affections is — will also die.

Saying all this hardly simplifies the problem. The computer is no better than its program; the structure of the questions put to it is the determining thing. So what appears to lie before us in a rather distasteful phrase is how, along with the quantifiable data, we can program the legitimate affections at work in a situation. If we can identify these affections and build them into the structure of the question we want to ask, then we can put the computer to work in our own interest. We can make it a brilliant extension of our own natural capacities simply by asking it the right questions fully and accurately stated.

This means, I think, that we must soon get heavily into the business of finding out what our genuine affections and intentions are. We cannot much longer resort to argument, or simple assertion, or name-calling, or the power play to get our own way at the moment. We will have to find out first what our own way ought to be so we can frame the right questions. This means, in the end, finding out who we are, and, more painful yet, accepting it. Then we can program our emotional necessities and desires and even, I would suppose, our aspirations. In other words, we can write satisfying programs for the computers when, and only when, we have learned to write satisfying programs for ourselves.

How do we get the information we need, and how do we make it explicit? This is not the place where you will get the final answer to a problem that has puzzled people since the beginning of recorded time. But some observations at least may be tentatively offered. Some of the information we need, some of the methods required to obtain further information are already available in the formal educational process. There is, to begin with, the old-fashioned source — the study of the humanities. Historically they have been an inefficient instrument, but they have always been and may remain, as Sigmund Freud said of women, "the best thing of their kind that we have." But I think we might spend more time now trying to figure out ways to increase their educative power. They serve too often as mere diversions or as objects for critical analysis. They should be approached in such a way that a student may be stirred by them, in such a way that he recovers his power, now almost lost, to be moved. The surest way to discover the existence and then to examine the meaning of the affections is first to feel them.

And then there are the newer means that give some promise, the social sciences like psychology, cultural anthropology, and the rest. They are still young and tangled up with too many long words and too much unorganized data. But they are today the most systematic methods we possess for analyzing the structure of the personality and for charting the course of the affections.

What the ultimate relation between the humanities and the social sciences will be I certainly don't know. I would hope that they would fortify each other, interact. I could make a case for instance that one could feel the full impact of *Hamlet,* be deeply moved by it, only after one had acquired an intellectual grasp of the structure of personality, the pattern of disturbed affections, so that one could recognize consciously what Shakespeare knew unconsciously.

One more thing about the uses of education here. It seems to me that we have reached a point of complication and sophistication in our culture where the things I have been talking about — the development of the instruments of industrial organization and our emotional and intellectual responses to them — cannot be learned once and for all in high school, college, graduate school, one Sloan Fellow year, or ten weeks in a senior executive development program. To live safely in our society, let alone manage it, will require a continuous education until a man dies. It is steadily necessary to learn more about the changing environment, about the novel instrumentations we are putting into it, and about ourselves as members of it. But education in the formal sense is not and never has been enough. We will need further incentive and means if we are to deal with our endeavor, which is to use the machine wisely by finding out who we are and abiding by what we find.

The primary incentive, and strangely enough perhaps one powerful means, lies in the computer itself. It is designed to give answers and to render decisions, but by its very presence and nature it puts the essential question, defines the central problem. Because it works only by simulation, it asks, it cannot avoid asking, of its maker, what is the nature of man? This is the oldest question. Put forward in the past, as a matter of philosophic speculation, it has been possible to avoid the full consequences of insufficient answers offered by honorable men or of false answers supplied in cunning passages by fools and knaves.

But the machine which simulates will take our replies today to use in immediate and practical applications. We withhold a full answer to the question it poses, we evade or equivocate at our own immediate peril. So the computer by its practical consequences can force man, for the first time, to raise up and examine his being (the being that must be, as they say, "programmed"), to face not only what he is but what he is doing, why he is doing it, and what he wants to do.

If history means anything, we will try to spare ourselves some of the pain of this investigation; we will, at times, get caught in traps sprung for us by fear or the pure intelligence. No doubt we are in for a dangerous time with an uncertain outcome. But there would seem to be no immovable obstacle to success. As I said earlier, much of our past difficulty has come because the problems have not been well defined. It should now be added that men have demonstrated impressive abilities to solve problems once the definitions have been reached. They have revealed fortitude in awaiting outcomes, intellectual enterprise in the selection and combination of data drawn from schooling and experience, imagination in supplying new values for unknowns in the equations set, continuing concern not only for what will work but also for what is lovely.

The computer has stated the problem bluntly. What is left is to refine the definition and to solve the problem. To the dangerous work of solving it there is no reason to believe we will not bring the same qualities we have possessed in the past. Since the machine can only simulate, the work of creating the future is still in man's hands. If he fails, if he misuses the machine, then in simulation it will misuse him in exactly the same way. Then, in terms of what we ever thought we could be or hoped to be, we would have had it — and for the last time. If we succeed, we will have acquired an incredible extension of the power of ourselves, our real selves.

5

A Little More on the Computer

I will make clear at the beginning two things that will, in any case, become obvious as I go along.* First, I do not know very much about the subject of these talks, the computer. And second, I have spent my life in that culture which, as C. P. Snow suggests, tends to produce nervous apprehension and depression of the spirit.

I do know that the computer in its present form is a relatively new machine; so I thought I might say a word or two as an historian on the way new machines and men have got on together in the past. I also know that the computer is a machine that will give answers to certain kinds of questions and supply solutions to certain kinds of problems. So I thought I might suggest what some men in my culture think they have found out about the perplexing dialogue between question and answer, problem and solution.

* At the first session of a series of discussions on the computer held at M.I.T. in 1961, C. P. Snow delivered the opening statement. This paper was a response to Snow's remarks.

As for the first topic, no more than Sir Charles am I a Luddite. One of the things you can learn from history is that men have lived with machinery at least as well as, and probably a good deal better than, they have yet learned to live with one another. Whenever a new device has been put into society — the loom, the internal-combustion engine, the electric generator — there have been temporary dislocations, confusions, and injustices. But over time men have learned to create new arrangements to fit the new conditions. Anything that has the power to build has also, of course, the power to destroy, and so, in the hands of men, do machines. But, on the whole and by and large and more often than not, men have always succeeded in organizing mechanical systems for constructive purposes and for the enlargement of human competence and opportunity. No one, I think, who compares the condition of life for the average person in the seventeenth century with the average condition of life today in our society can fail to reach this conclusion.

Partly for this reason I am not, in the matter of Sir Charles's specific apprehensions, as much of a Luddite, I guess, as he may be. Take his first apprehension, that the computer may measurably increase the tendency toward closed decisions in our society. Obviously we will have to think about this. Machines can, beyond doubt, alter some of our views of things; the multi-engine plane, for instance, has changed somewhat our sense of time and space. But there is, as I understand it, nothing in the nature of the computer that will necessarily take us nearer to closed decisions, closed decisions such as those taken in the days of Wolsey or Richelieu or Caesar long before there were radar sets or computers. Both the machine and its programmers will have to work within a general scheme, a field of general decisions and determinations that can still, as Sir Charles says, be gathered out of the air if that is the way we want to do it. In determining the kind of life you want to have, the instrumentation is less influential than the nature of the culture you create to control what you want to use the instruments for.

For example, I do not believe that the rumble seat of automobiles increased the incidence, it merely changed the locus, of experiment in physical relations between boys and girls in the age of F. Scott Fitzgerald.

Then there is the apprehension about the computer as a fascinating gadget. It is obvious that there is always danger from the gadget-happy, whether the gadget is a machine, an idea, or a procedure. Amasa Stone, for instance, a very able man, killed a trainload of people because, against advice, he built a bridge at Ashtabula from a truss design for which he had an ancient attachment. I myself, for example, have no tonsils, nor have my brothers, nor have most of my generation, because an accomplished ear, eye, nose, and throat man and his colleagues were all obsessed with the thought that the way to make a boy grow was to take out his tonsils.

In an age of new departures we have to live with all this, I suppose. But only, in each case, for a limited period, so history suggests. Over time the potentials of a new gadget are explored by trial and error until the real capacities are discovered and understood. Then, whatever it may be — Manichean heresy, steam turbine, penicillin — it is fitted into a reasonable context.

I do not want to appear like a wise old head, made sager by my study of history than those like Sir Charles who have actually been there. Of course what he has said should cause any sensible man to think, and what I have said does not, I know, fully dispose of the complex problems he raises. Perhaps we can all talk about these things together afterward. But now, I haven't much time, and I want to get on to a nervous apprehension of my own.

I think we may have more difficulty in exploring the full limits of the computer than we have had with earlier gadgets. I think there may be more danger in the period of trial and error than there has been with earlier devices. These earlier devices — looms, engines, generators — resisted at critical points human ignorance and stupidity. Overloaded, abused, they stopped work, stalled, broke down, blew up, and there was the end of it. Thus they set clear limits to man's ineptitudes. For the computer the limits, I believe, are not so obvious. Used in ignorance or stupidity, asked a foolish question, it does not collapse; it goes on to answer a fool according to his folly. And the questioner being a fool will go on to act on the reply.

This at least is what my culture tells me often happens. Let me give you an example. In the play with which you are all familiar Hamlet had a problem which he defined for himself as follows: What happened to the late King of Denmark, and what should he, Hamlet, do about it? Framing the question accurately — a good program — he took it to a ghost, the most sophisticated mechanism in the late sixteenth century for giving answers to hard questions. From the ghost he got back a very detailed reply which included a recommendation for a specific course of action. Responding to these advices, Hamlet created a political, social, moral, and administrative mess that was simply hair-raising.

The trouble was that he had got the right answer, the answer he deserved, to a question that was totally wrong. He had asked about his father when he should have asked, as any psychologist will tell you, about himself and his relations with his mother.

My culture says, in other words, that it is much harder to ask the right question than to find the right answer and even the right answer to the wrong question isn't worth much.

Some of you, like some of my students, may say that *Hamlet* is only a play, so what does it prove? So I will give you some further evidence about questions and answers taken from real history. We once asked how we could limit the arms race between us and Britain and worked out the answer that for every British light cruiser we could have 1.4 American heavy cruisers. About the same time we asked ourselves how to make the nation "self-sustaining" and arrived at the answer of the Smoot-Hawley tariff, which set an average *ad valorem* rate of 40.1% for all schedules.

You will know, I am sure, that these answers caused us very real trouble of one kind or another. They did so because the questions they were designed to answer were framed in a wrong interpretation of events, a false conception of the actual problem. The answers supplied therefore gave the wrong solutions. They represented collectively what Ramsay MacDonald said of the cruiser ratio: "an attempt to clothe unreality in the garb of mathematical reality."

The quotation from the Prime Minister suggests a further source of nervous apprehension, the tendency to simplify human situations and to do so, often enough, by reducing them to quantifiable elements. I have spoken of Hamlet, so, by way of illustration, I will speak of him again. I remember two things my Tech students have said in explanation of his behavior. First, he had too much feedback in his circuits, and second, he was 16 2/3% efficient — because he had one person to kill and he killed six. This, purely incidentally, is about the thermal efficiency of the average internal-combustion engine.

What I want to suggest here is the persistent human temptation to make life more explicable by making it more calculable; to put experience into some logical scheme that by its order and niceness will make what happens seem more understandable, analysis more bearable, decision simpler. When you talk of 1.4% of a cruiser you can hope you have solved the underlying diplomatic issue you haven't dared to raise; when you pass a tariff with average rates of 40.1% *ad valorem* to make a nation self-sustaining, you can assume you don't have to look further for the causes of the worst depression in the nation's history — to which, incidentally, you have just contributed by passing the tariff at all. This is, I suppose, the way it does figure; and this seems to have been the human tendency from the time of Plato's quantification of the Guardian's role right on down.

I am not trying to suggest that the computer will soon bring us all under the cloak of the mathematical reality of its programs. But today the tendency to work with quantifiable elements and logical systems seems to me accelerating. There are more tests and measurements (the brain of a candidate for college works within a precisely graded scale from, presumably, 1 to 800), more rational systems like those of Keynes and Freud to assist us in ordering the economy and the personality, more mathematical models, and more efforts, as in the schools of management, to reduce administrative experience to quantifiable elements. This, in the name of clarification and the advancement of general understanding, is quite obviously all to the good. The aim of pure reason, which proceeds upon measurable quantities, is, presumably, to introduce increasing order and system into the randomness of life. But I have here the apprehension that as time goes by we may begin to lose somewhat our sense of the significance of the qualitative elements in a situation, such things as the loyalties, memories, affections, and feelings men bring to any situation, things which make situations more messy but, for men, more real. My apprehension is that the computer which feeds on quantifiable data may give too much aid and comfort to those who think you can learn all the important things in life by breaking experience down into its measurable parts.

I hope it is clear that I am for order and logic. But I do hope also that it remains clear that in all the really interesting questions and problems of life the measurable and the immeasurable are all mixed up. I think from time to time of the Pythagoreans, those men who came to believe that "all things are numbers" and were supposed to have put to death a man who suggested the idea of the incommensurable. Even in an inventory program the risk one is willing to bear if he runs out of stock must be considered, and this risk is in part determined by the unknown size of the irritation of frustrated customers. Even in the strategic bombing exercise suggested by Sir Charles there is what might be called the Coventry factor to put in the program, the unrequited feelings of men and women who had been bombed with no redress.

Still, I am no Luddite. What I want to do, first, is to find out all there is to find out about the computer. And I must say this is hard enough. In my cursory researches I have been told a great many different things by people who have at least thought more about it than I: that it is and always will be simply an idiot doing what it's told to do; that it can now design fractional horsepower motors and electrical circuits; that it can already throw old data into new combinations — introduce intellectual surprise (and this, said the tired pedagogue, is rarer than you think); that it may sometime write a sonata; and that, if we can only get enough vacuum tubes (ten to the tenth is the figure) hooked up right, we cannot exclude the possibility that it may feel its own emotion and have a will of its own. The spread between assumed present capacity and the foreboded potential is, in other words, considerable.

In any case it is here, and no doubt it is going to develop. Everybody, or almost everybody, seems a little uneasy about this, and why not. This is man's first encounter outside himself with something that is exactly like some inside part of himself. It is not, as many other machines have been, like his arms or hands or legs in the work it does; it's like him. How much like him we don't yet know. But regardless of what happens in the future we have already made a machine that simulates some part, a small part, of what we alone have been able to do in the past by taking thought. Even this small advance begins to raise the large question we have succeeded so often in avoiding — as Hamlet did for instance. What is in fact our true image, what is our real likeness?

To assist us in exploring this subject further, I propose, not that we bust up the machine, but explore it in a series of experiments suggested to me by Sir Charles's idea of replaying the bombing problem.

I suggest a continuing experiment in which the machine would be asked to reconstrue a series of situations out of the past, situations taken from my own culture, in which men have acted most successfully on their own. One could begin with simple things — a successful plant relocation — and work up through, let us say, the Army and Navy decision on how to use the plane against the submarine, and then on to the most interesting situations — our method of limiting trusts and cartels as revealed in the Sherman Act and its subsequent modifying judicial interpretations, or the miracle of Queen Elizabeth I's foreign policy. I suggest for each situation a series of experimental programs, leaving some things out, adding some things, practicing ways to code things that seem uncodable and so forth, to test various hypotheses and understandings by repeated trials against what may be called the real situation.

Some of the problems may be, at the moment, a bit far-out, given the present state of the art. Also it is probable that the study of a particular past and nonrecurring situation would not produce the most desirable results. For instance, it has been suggested that a program designed from evidence taken from the record of six or eight different revolutions might well be useful in the attempt to discover how to control common elements, both the stabilizing and the disrupting, that operate in all revolutions including the one we are in today. Or a program put to a computer on the problems of underdeveloped regions that are now developing under the energy of technology might well include data collected from the history of other regions that have passed through this stage of experience.

Whatever difficulties or defects there may be in these particular proposals, I'm told a beginning of some sort could be made in the direction I suggest, and I hope it will be, for several reasons.

First, a good many people would have to learn some history, which is a good thing.

Second, it would be a way to capitalize on the ancient truth that fools persisting in their folly may learn wisdom. We could acquire understanding of the uses of the machine without danger in a time of controlled trial and error.

Third, the machine would therefore become not so much a problem solver as a learning machine, which is today in fashion. Used as suggested, it would force us, again without danger, behind the silly or distracting questions we like to ask to the real questions we have to ask, and teach us to ask them more correctly. It would thus help us to sort out the things that could be thought from the things that can only be felt, and advance a little of our understanding of how much feeling goes into what we call thinking. Perhaps some of the things we feel may turn out to be more identifiable and explicitly definable — as the work of Freud indeed suggests — than we think. But if not, if they can't be programmed as they say, at least by this exercise we may well find out more about their meaning and influence. What I am suggesting here I trust is obvious; that we use the machine that only simulates to explore fully what it is simulating — what image, what likeness.

For some this assault by mechanism upon what e. e. cummings calls the "single secret that is still man" will be distasteful and to some appalling. It is not, at least not to me. Over on the other side good friends of mine are using accelerators to find out the secrets of other marvelous structures. The more they find, the more they seem to stand with respect, indeed with wonder, before their findings. If man working experimentally can learn from something he invented more about himself, who he is and who he isn't, then he will have learned enough, I should think, to use the machine that simulates and also himself more wisely and constructively.

And if in this process he finds, at length, even the full answer to the single secret — well, that's what my crowd has been trying to do for 2500 years.

6

Men and Machinery

This is an account of what happened in the United States Navy when a new ship was put into service in the middle of the last century.* My remarks are cast in the form of a small anecdote interrupted twice by extended digressions.

The ship was the *Wampanoag*. Laid down at the Brooklyn Navy Yard in 1863, she was commissioned in the naval service five years later in 1868. She measured 355 feet on the water line; her beam was 45.2 feet; she displaced 4200 tons. For her time she was a big ship. Like all vessels of that day, she had masts and sails, but unlike other vessels her primary source of forward motion was her steam propulsion plant. She was driven by an engine with two cylinders 100 inches in diameter and with a four-foot stroke. The power in her engines was transferred through wooden gear wheels to the shafting that turned a propeller 19 feet in diameter. She carried a 60-pound gun, two 100-pound

* This essay was delivered as one of three lectures at the California Institute of Technology in 1963.

guns, ten 8-inch guns, and four small howitzers. For her time she was heavily armed.

This vessel went on her sea trials in February 1868. Beginning at nine o'clock on the evening of February 11 her captain put her through her tests at full steaming power. This trial continued for thirty-eight hours in seas that were for the most part heavy. Over the full period she maintained an average speed of 16.6 knots, or 19.14 statute miles. At one time, for six consecutive hours, she steamed at 17.25 knots, or 19.89 statute miles an hour, and for one hour, she reached and maintained a speed of 17.75 knots, or 20.465 statute miles per hour.

When the *Wampanoag* completed all her sea trials, a board of engineers gave the opinion that her records could not be equalled "for speed or economy by that of any sea-going screw vessel of either the merchant or the naval service of any country." Her commanding officer in his official report said, "I consider the *Wampanoag* as a ship to be faultless in her model and, as a steamship, the fastest in the world." There was no doubt about this. The fastest ship afloat at the time was a merchantman, the *Adriatic,* which once had run a measured mile in smooth water at a speed of 15 knots. In fact, in the next two decades no ship was built that could match the performance of the *Wampanoag*. Half a century after her trials, the great marine engineer George W. Melville gave it as his opinion that "she was a magnificent success in every way — really in many ways the greatest success as a steam war vessel that the world has ever known."

The *Wampanoag* in other words was quite a ship. There are so many interesting things to say about her design, her building, and her career in the Navy that I hardly know where to begin or how to proceed in the time allowed. My plan is to select only two or three things of perhaps special interest out of her history and to add, as we go along, some reflections of my own. I would like to start with some biographical remarks about the man who conceived of, designed, and built her.

His name was Benjamin Franklin Isherwood. He was born, the son of a New York City physician, in 1822. At the age of nine he was sent off to school at Albany Academy, where he remained for five years until he was sent home for what was called "serious misconduct." My study of invention in the nineteenth century leads me to conclude that such expulsion was often one of the most important parts of the formal education of the men who did great things in the technology of that marvelous period. And if it wasn't school that got them into trouble, it was often enough money, or alcohol, or women, or their own troubled selves. Make of it what you will — a sense of frustration with the conventional arrangements, boredom with things as they are, hostility to powerful fathers, rebellion against pious mothers, and all that — something apparently pushed them to surprise or irritate or appall or just to change their surroundings. Today this is often called creativity, and a great search is on to find out how it can be made into something the normal, fully adjusted man can do in any well-ordered corporation. At any rate, Isherwood got off on the right foot by getting expelled from Albany Academy at the age of fourteen.

He then took up work in the shops of a railroad that ran from Utica to Schenectady. Here he labored under a master mechanic named David Mathews. Then in due course, still in his teens, he joined his stepfather, David Green, who was in charge of the construction of the Croton Aqueduct. After this he returned to railroads — to the Erie — again in shops and once again under the supervision of a remarkable engineer, Charles B. Stuart. From there he went on to building lighthouses for the Treasury Department. This led him to an interest in light and optics. He designed some lenses and then went off to France to take part in their manufacture.

Returning to this country, he went to work in the Novelty Iron Works in New York, which dealt with the fabrication of all kinds of iron members for metal structures. There his old supervisor Charles Stuart, who had just been appointed Engineer in Chief of the United States Navy, sought him out in 1844 and offered him a commission in the Corps of Naval Engineers.

For the next few years Isherwood went to sea in the first steam vessels of the Navy. There were not many of these ships, six or seven, and none had been designed as steamers. They were in fact old sailing vessels to which auxiliary engines had been added. These engines were small and primitive in design. Isherwood's job was to keep them running, and since they were always breaking down, he learned a lot. During this period he also learned a lot about the way the engine power could be applied to drive a ship. It was a time of great argument over the relative virtues of different kinds of gearing mechanisms and, more particularly, over the relative virtues of water wheels and propellers. Isherwood designed the first feathered wheels, but he was especially interested in the development of more efficient propellers.

It was, in fact, a time of great argument and indecision about everything which had to do with the steam propulsion of ships at sea. One of the largest areas of uncertainty involved the proper point for the cutoff of steam in the cylinder. There was a school of thought that held that Boyle's — or Mariotte's — Law on the expansion of gases obtained also for steam in a cylinder. The theory was that you could cut off the steam early — put only a little in the cylinder — and let the power in the steam's expansiveness do most of the work. There were obvious economic advantages in this procedure which attracted many supporters. But a few others — in the forties and fifties, a very few — held a different opinion and argued for a much longer cutoff point. Isherwood entered the discussion of this matter at a time when the discourse was stabilized at a level of most sordid dialectic. He was a long cutoff man, and for his opinion, he was called an "ignorant child," a "merchant of pure nonsense," and a peddler of the "manifestly absurd."

It occurred to him that he might establish his point by actual demonstration rather than by resort to the *argumentum ad hominem.* Accordingly he began in 1856 a series of experiments on naval vessels laid up in ordinary. His most famous work was done on the U.S.S. *Michigan* in the year 1860. These were really elegant experiments. From them he learned a great deal about the nature of steam and about the mechanical systems designed to put the energy in steam to doing work. Among other things he demonstrated that the presumed advantages of using steam expansively were not nearly as great as his opponents claimed. He also learned a great deal about cylinder sizes, valve design, and boiler capacity. Among the major conclusions he drew from all the data produced was this one: the simplification of steam engines is of the first consequence to both success and cheapness. Interestingly enough this small conclusion annoyed those opposed to him almost as much as did his demonstration that the short cutoff point was a profound mistake. Working on the whole with uninformed intuitive perceptions, most engineers at the time were trying to control the novel energy they did not fully understand by creating elaborate and complicated mechanical systems. I have seen this happen in other fields including, I suppose, my own — the smaller the understanding of the situation, the more pretentious, often, the form of expression.

What Isherwood found out by his four years of experiment, both his evidence and his conclusions, he published in two great volumes called *Experimental Researches in Steam Engineering.* These became a basic source of information for the design of steam power plants for several decades. His findings also became the basis for his own designs for the engines of ships in the United States Navy. In 1861 he was appointed Engineer in Chief of the service. At that time there were ninety ships on the Navy list, of which twenty-one had steam power. In the next four years Isherwood personally directed the design and construction of the power plants for the six hundred steamers added to the fleet in the course of the Civil War. His machinery was remarkable for its simplicity, efficiency, and durability. What he accomplished in the face of extraordinary obstacles, technical and political, was later recognized as a matchless achievement. At the time he labored, as he said, "under great difficulties and discouragements — moral and physical — additional to those normally belonging to the deranged state of the mechanical art at that period."

I have spent so much time on Isherwood for several reasons. First, the things he did in the years from 1840 to 1860 had much to do with the things he did in the design of the *Wampanoag*. Second, his career is a nice, small footnote to one of the interesting and, on the whole, neglected passages in American history. This passage has to do with the way we accumulated the intellectual capital that supported our remarkable industrial development after the Civil War. Many of the men who took part in that development, who, in fact, built the plant, were trained in the years before the war when there was no R.P.I., no Stevens, no

Case, no M.I.T., no C.I.T. These men — people like Holley and Fritz in steel, Westinghouse and Shallenberger in electricity, Roebling and Stuart in civil engineering — learned their arts and trades on the job, in small shops, local foundries, and little ironworks. There was also the government service: Coastal Survey, Hydrographic Office, Lighthouse Service, and the Corps of Army Engineers. And also, and most important, the great technical educational institutions: all those small railroads and canals. The men who went to school in these places grew up in a changing curriculum, within a simple but evolving technology, and by their works, contributed to the acceleration of that evolving process. Starting often by acquiring skill in a single element — puddling, forging, casting — they were soon forced into thinking about how to bring together a whole set of related elements. For instance, the gradual assembly of the several parts of the integrated three-high rail mill. And by this process they contributed to the integration of the larger technical system that became industrial America.

By a review of such considerations I am brought to the third reason I have spent so much time on Isherwood's career. I do not really know whether one learns more and better and faster by proceeding from the particular out to the general proposition or by mastering the general proposition and then seeking to apply it in the particular situation. I suppose it depends a lot on the subject and the man. But in any case, I am much struck by the character of Isherwood's early instruction. He began, as I have said, making many different kinds of small things and repairing or reworking many different parts of big things, like locomotives. He did all this under the direct supervision, as an apprentice really, of three different skilled men, master mechanics. By these experiences he got the feel of different kinds of materials and the feel of a variety of mechanical problems. He also got instruction by example from the masters in the elements of individual style.

As he proceeded further in his experience, pushing into more elaborate situations where rules of thumb, or unsupported intuitive perception, or ancient practice would not serve, he became aware of the need for more sophisticated understanding of his work. Accordingly, he designed those elegant experiments with the machinery of the naval vessels laid up in ordinary. From the data he thus obtained he arrived at general propositions concerning the properties and behavior of steam, at what one of his opponents called "*mere hypotheses* of his own that he had printed in a *book*." Governed in his thought by these hypotheses, he designed the best marine steam power plants that there were in the world.

All this seems to me to be the perfect progression for the intellectual development of an engineer. If I understand this matter correctly, engineers must begin with the urge to apply, that is, the urge to make things, that is, the urge to take separate, diverse elements and put them together into an organized situation or construct that will really work. This is the urge, if I understand it, of the artist, to give some concrete form to an organizing perception. This urge seems difficult, really impossible, to acquire by training. Either you have it or you don't. And, it turns out, it is an urge that is hard to educate by any formal institutional means that we have so far devised. Rarely, for instance, do our novelists, poets, musicians, and painters come out of colleges or universities. These are better places to exercise the cerebral cortex than the primal urges, better places to sharpen the powers of investigation and analysis than to stimulate the artistic impulse.

Such impulse classically has often been better served by

actual practice under the true conditions of apprenticeship. This is the kind of education Isherwood had. But he went on, as I have said, from simply making things into a concern for theoretical structure. Indeed, there is to be found throughout his career a most satisfying balance between practice and theory.

This balance has been, for the most part, precariously maintained in the intellectual atmosphere of this country and, perhaps, in the intellectual atmosphere of any country. It is therefore worth noticing how Isherwood achieved this satisfying equilibrium. Starting, as the times required, with practice, work, making things, his continued interest in such operations appears to have determined the limits set upon his theoretical cogitations. The structure of theory he created for himself was never bigger than it had to be to give him a better sense of what he was doing and a sense of how to do it better.

The kind of education Isherwood received is now hard to come by. The small shops and local canals are difficult to find. The problems posed by ordinary machines, automobiles, generators, and electric motors, are often easily solved by reference to the manufacturer's handbook and the spare-part bin. More important, much of our machinery is now an extraordinary, novel, and complex expression of the findings of rapidly developing science. So the tendency in our engineering education is, increasingly, to begin with theoretical considerations, a solid grounding in the physical sciences and mathematics. This, of course, in many ways is all to the good. But under such conditions the simple urge to make things, and all those intuitive perceptions required to bring form and function nicely together in the making of things, may wither away. They seem to me at times to wither faster now that science has become apparently so much more, intellectually, professionally, and even socially, the correct thing to do.

I am raising in a somewhat different way, I find, the point I tried to make earlier in this essay. I have an uneasy feeling that in engineering, as in other parts of our education and perhaps in many other parts of our experience, we are getting out of touch with the single, limiting circumstance, the resistant, intractable material, the hard particular that gets snowed by the general theoretical proposition. Practically all of the students I teach for instance, who come to us from the best colleges and universities in this country, have the vocabulary to express the general ideas they have been taught. They know all about input and output, feedback, the Oedipal situation, subdominant profiles, the uncertainty principle, macro and micro, charismatic personality structure, and all that. But very few, if any, can put a simple thought of their own into a single simple English sentence that will express that thought.

But back to Isherwood and the *Wampanoag*. Isherwood, you will remember, became Engineer in Chief of the Navy in 1861. At the start of his administration his primary concern was to put engines in as many existing vessels as possible to strengthen the Northern blockade fleet. He performed this task brilliantly, but at first his influence on any larger aspects of ship design remained small. As Admiral Sampson had said, "sail [is] the primary motive power, it [is] considered a sufficient concession to admit steam on any terms." But as the war continued, certain influences developed that caused Isherwood and others to go beyond this rather limited frame of reference.

One of these influences was the success of the Confederate commerce destroyer *Alabama*. The dramatic destruction of Union merchant vessels produced by her independent action greatly impressed all observers. The independent action was, in turn, produced in part by the imaginative daring of her officers and crew and in far larger part by her speed. Though she made only 9 knots, she was in fact superior to almost all other existing vessels. This fact caused the Department of the Navy to think. Obviously a fast, heavily armed ship could do great damage to the commerce of a nation at war. And, in the middle of the Civil War, their departmental cogitations were further stimulated by conditions abroad.

In 1863 word came that France was trying to persuade England to intervene with her in our rebellion. That England would act seemed doubtful but still possible; and the means to respond to her possible action were not readily at hand. In these circumstances Isherwood proposed to the Secretary of the Navy that a class of ship should be built that would directly threaten England in her most vulnerable point. That point, he argued, was not the British fleet but the British trade, and the *Alabama* had demonstrated what was needed to disrupt trade. The plan was to build vessels of great speed, of superior firepower, and of marked ability to keep the sea for a long time. These vessels were to follow the principal tracks of commerce and to destroy every British cargo carrier they found. As Isherwood said, "they were to be built for business and not for glory. They were solely to attack the enemy's purse, and to bring him to tears of repentence in that most tender point."

Such arguments carried the day. In 1863 contracts were let for four such vessels to be built in different yards by different men. Since the idea was novel, Secretary of the Navy Gideon Welles wished to proceed in the mood of experiment. Isherwood was asked to build two ships in government yards. John Ericson was assigned a contract for a third in a private yard, and E. N. Dickerson assumed responsibility for a fourth, also to be built in a private yard.

John Ericson was a great man, the hero of the *Monitor*, an experienced, imaginative engineer. Dickerson was a picturesque figure who had already built three elaborate and dramatic failures for the Navy. He believed in the short cutoff and the power of steam used expansively. He was, among other things, a lawyer who put forward his engineering theories in long letters to the Secretary of the Navy, in communications to Congress, and in articles to the newspapers. He was apt to fortify his most technical arguments with long quotations from the lady who did his laundry; and once in court he interrupted his summing up of a case before a torpid jury to give an extended speech on "The Navy of the United States. An exposure of its Condition and the *Causes of its Failures*." The prime causes in his opinion were B. F. Isherwood and the long cutoff point.

The four vessels were laid down in 1863. I wish I could go into the differences in the design of each in some detail. But I will confine myself to some brief remarks on the work of Isherwood. He began first not to think about the engines alone, but about the engines in relation to the hull. This was a novel idea. Up to that time you simply put an engine in a conventional sailing bottom, as later you simply put an internal-combustion engine in a carriage or designed the first iron bridges along the old wooden truss lines. Apparently, man only trusts his imagination so far and seeks to put his new notions into the comforting forms of familiar structures. In any event, Isherwood broke with existing practice in two ways. He first laid out his power system: boilers, engines, gearing, and shafting, and then, as he said, "sort of built the hull around the whole interior arrangement." But he had laid out his power system with the idea that the hull must be built for speed, so his encasing hull was narrow beamed, fine bowed, long, and sleek, a really lovely vessel.

He also devised a novel and ingenious way to transfer power from the engines to the propeller. He feared that the conventional system of connecting rods would produce such vibration in the narrow hull that the ship would shake itself apart. So he constructed a new gearing arrangement which was one of the principal reasons the *Wampanoag* proved superior to the ship designed by Ericson.

The four vessels begun in 1863 were unfinished when the war ended in 1865, and there was much discussion over whether the construction should be continued. Economy was one reason put forward, but another was that the ships were not worth finishing because they were obviously failures. The two built by Isherwood were the chief targets of attack. As an editorial in a service paper said, "He is ruining the Navy by his untenable steam delusions. . . . If we have utterly condemned his theories we have done so only on a basis of irrefragable facts. . . . We are free to say that there is no such monument of mechanical incapacity as the steam machinery of the *Wampanoag* class to be found in the annals of marine engineering."

In spite of such harsh judgements, the work on the vessels was continued by order of the Secretary of the Navy. They were all finished in late 1867 or early 1868. Mr. Dickerson's *Idaho* was first in the water. She broke down so often that her sea trials could not be finished and maintained a speed of only 8.27 knots in one twenty-four-hour period. The trial board reported that the ship "had not in any particular been fitted and equipped in accordance with the obligations" and recommended that she should be "condemned and rejected as totally unfit for service in the Navy."

Then came Mr. Ericson's *Madawaska.* She made a more favorable impression, but proved difficult to handle in a seaway and never came within 2 knots of her rated speed of 15 knots. Finally in February 1868 the *Wampanoag* put to sea with the results I reported at the beginning. She handled beautifully in any weather, never needed more than two men at the wheel, and steamed with great economy at 3 knots faster than her designed speed. She was, as already reported, the most remarkable man-of-war in the world, and she put the United States Navy a full generation ahead of all other navies.

I find I cannot resist the temptation for another digression. Some years ago my friend Edwin Land, something of a hero himself, came to M.I.T. and delivered a speech in which he spoke of the difficulty of finding heroes in the engineering profession. The law had its Blackstones, Cokes, and Holmeses, medicine its Cushings and Oslers, science its Pasteurs, Boyles, Maxwells, and Einsteins. His point was that engineers were deprived of the great figure as a standard to repair to. This is not a small point. I remember Aaron Copland explaining how difficult it was for an American composer seventy or eighty years ago to write what he called serious American music. He had in mind Edward MacDowell. Had MacDowell been born in Europe, he said, he could think of writing a symphony like Schumann or Mendelssohn or Brahms. Here in 1890 he could write an American symphony — like whom?

In engineering there are irr fact some heroes: Edes and Marconi and Isambard Kingdom Brunel. But it is true, they seem few and far between. The reason, I think, is not so much that the field does not attract the hero, the great man, as that the work itself so often has strict limits set upon it. Precedents in the law, formulas in chemistry, equations in physics, relationships in physiology once developed tend to go on having life and continuing influence. The particular solutions of engineers are on the whole local, limited by time and place and singularity. And the rate of change in instrumentation makes many constructions tend toward rapid obsolescence. The artifact moves far more rapidly toward extinction than an idea does. Yet the work itself and the men who have done it have often reached heroic proportions.

Take the case in question: of the forgotten Isherwood. There are the outward conditions of heroism — great difficulties in the task itself, bitter opposition, public ridicule — all calmly accepted and worked through to impressive success. There is also the less dramatic but more impressive set of circumstances that produce what may be called intellectual or professional heroism. Isherwood took a single element, the marine steam engine, which in its primitive condition was surrounded by ignorance. By his work he moved far beyond the existing state of the art. This in itself was a considerable achievement, but he went beyond that. He thought of the effect of this single element upon all the other connected elements. Because he did so he went on to develop novel ways to transfer the power available in the engines and then proceeded to act with success in a related field not strictly his, ship design. He then went even further in his calculations. Proceeding from his concept of the fast, heavily gunned vessel that could keep the sea for long periods, he reached the conclusion that such ships if built in large number would alter the contemporary ideas of naval warfare. Rejecting the existing doctrine of harbor defense, blockade, and independent actions, he produced the concept of total command of the sea by the actions of fleets of ships in remote waters. He argued this proposition with force and clarity thirty years before it occurred to Alfred Thayer Mahan at a time when Mahan had all the material elements at hand to force him into making such a generalization. Such transformations, not only in the instrument itself but in the understanding of the more general implications of the way the instrument could be used, seem to me the product of what may be called the heroic view.

Isherwood was favored of course by the times. He flourished in a period when much that is now standard practice, simple problems for the computer, constituted expeditions into the dangerous unknown. Charles Stuart, the man who had brought Isherwood into the Navy, made his reputation building a bridge over the Niagara River that most qualified observers thought would not hold a train of cars right up to the moment the cars were put on it. It was also a time when an engineer could by himself deal with an entire system, a whole bridge, a whole engine, a whole power plant, and so on. He was thus in a way forced out of the particular into more general considerations and the heroic proportion.

Not so today, at least not so often. We have arranged new ways to solve our complicated problems, dividing them up into small constituent parts and farming the parts out to men with special competence in limited areas. This is not only true of engineering. It may even be happening or about to happen in the intellectual field where many of our most recent heroes came from. The conditions that produced a Bohr, a Dirac, a Pauli are changing. There is not so much possibility of *ein Heldenleben* on the neutrino night shift, for example.

Whether or not this is a bad thing, I will not examine here. But having spent so much of my time in thinking about the nineteenth century, it appears to me that we could think a little about inventing ways to solve our problems, technical and other, in a manner that would permit men to consider more than they often can now, the whole instead of the small part, and to consider, or to see, the relation of the several small parts to the whole and the relation of the whole to other things.

From this digression I now return for the last time to U.S.S. *Wampanoag*. For one year after her sea trials she remained in the Atlantic pursuing her lawful occasions. In 1869 she fell under the scrutiny of a board of naval officers appointed by the Secretary of the Navy to report on all the steam machinery afloat at different Navy Yards.

The general mood in which the members began their work is well described in U.S. Document 1411 of the 41st Cong., 2d Sess. (1869-70), Volume One, Part One. The steam vessel, so the board reported, was not a school of seamanship for officers or men. "Lounging through the watches of a steamer, or acting as firemen and coal heavers, will not produce in a seaman that combination of boldness, strength and skill which characterized the American sailor of an elder day; and the habitual exercise by an officer, of a command, the execution of which is not under his own eye, is a poor substitute for the school of observation, promptness and command found only on the deck of a sailing vessel."

With such a philosophical predisposition, the board set out to examine the *Wampanoag*. They found, in principal part, as follows: the area of the grate surface under the boilers was greater than the area of the immersed midsection. Since ordinary practice was to fix the proportion the other way, in a ratio of 1 to 1.63, the board was brought to the conclusion that, as they said, "it was very remarkable." Second: her length would probably produce "a palpable want of longitudinal rigidity" with, as the board said, "a resulting effect that would at least prove a source of perplexity." Third: because of her length she would "present an immense target to an antagonist," especially, the board added in fuller explanation, "in a lateral sense." Fourth: "It was fair to infer," the board said, "that a vessel of [this] sort is liable to much rolling," and it was also fair to infer that "a surging motion, in a seaway, must ensue." Fifth: any vessel of such large displacement produced a "difficulty in maneuvering either to gain or maintain a position." Sixth: the placing of her spars and masts nullified the use of canvas independently to any advantage of moment over other cruisers that had sails.

This was the principal bill of particulars against the *Wampanoag*. From these particulars the board drew the following general conclusion. The ship was, it said, "a sad and signal failure, and utterly unfit to be retained in the service," and therefore she would prove "a happy riddance." The fact was that no changes would improve her; she was "too much of an abortion."

The most interesting thing about this report is that all the things the board thought were perplexing, or likely to happen, or that appeared in the light of earlier experience palpable, or that were very remarkable, or that seemed fair to infer — all these were things that in actual practice had not turned out as it seemed fair to infer and presume. The ship's trials had revealed that the ship handled well in heavy weather, that she turned and maneuvered rapidly, that, as two special observers of the Secretary of the Navy reported, she was in fact remarkably "steady," "efficient," and "easy."

As a result of the report of the Naval Board in 1869, the *Wampanoag* (and the other ship built by Isherwood as her sister, the *Amonoosuc*, which performed as well) was laid up in ordinary for a year. Then the *Wampanoag* became a naval receiving ship at New London, and her name was changed to the *Florida*. Then, some years later, she was sold out of the Navy.

Now it must be obvious that the members of this Naval Board were stupid. They had, on its technical merits, a bad case, and they made it worse by the way they tried to argue it. They were not better than those other officers I spoke hard words about thirteen years ago who had opposed a way of firing a naval gun more accurately because it was a new way. My conclusion thirteen years ago was that in a world such as ours, new ways to do things were standard operating procedure and that we had all better realize it and become an adaptive society before we were shaken apart or disintegrated under the strain produced by our blind resistances. The message was to live like Elizabethans or Italians of the Renaissance, to find a new salvation "in identifying ourselves with the adaptive process" and find "security, satisfaction, joy and exuberance in going out to meet the changing times."

My point then was really quite a simple one: the officers who opposed the new system of gunnery did so because they were victims of fixed attitudes, tied to a set of feelings built up around a different system of gunnery. My initial object this time was to repeat my previous success simply by repeating my previous talk. By cleverly changing the instrument from a gun to a ship I would drive home the point that old sailors never learn; they're just culture-bound.

To do this now, all I have to do is explain a little more about the culture. In the first place, all the officers on the board had grown up in sail in days when the intrusion of steam was a concession. There was also, beyond this immediate professional environment, the general cultural surround, what might be called the conditions of the times.

Some of these conditions were admirably described by two officers who thought well of the *Wampanoag*. However, they explained to the Secretary of the Navy, in a separate report, that she ought to be greatly modified "inasmuch as the main and special purpose for which this vessel was built no longer exists for a navy in time of peace."

Peace in those days had three dimensions; it was an actual condition. For instance, about this time the chief officers of the Navy reported to the Secretary of the Navy in italics that ironclads *"are absolutely needed for the defence of the country in time of war,"* but they said they opposed the building of these absolute necessities since in peace "this weakness of the country should not give rise to unnecessary alarm." There were in those days a good many other reasons for building men-of-war, and these officers made the reasons very clear. They said, "It is our opinion that, owing to the large supply of suitable timber at present on hand in Navy Yards, which the interests of economy demand should be utilized, the familiarity of our eastern workmen with wooden ship building, and their dependence on it for a livelihood, the resources of the country with respect to this material, and the possibility of building wooden vessels of a limited size that shall be staunch, efficient and economical, the 10 kt class of vessels should be built of live oak frames, planked with yellow pine."

It was at about this point in my researches that I began to be aware of a growing sense of dis-ease. I began to see that it was not just that the failure of the *Wampanoag* could be inferred in theory in spite of her demonstrated practical success. There were other factors. Since there was no warship like her, there was no natural opponent for her, and since she was built for war and the world was at peace, there was nothing for her to do. And the decision of the Naval Board that condemned her acquires additional weight when observed in the light of subsequent events. For a good many years after her sea trials, far beyond her normal life expectancy, no navy did produce a vessel that equalled her in speed, and also, for a good many years there would have been nothing for her to do. The world remained, as the officers predicted it would, at peace. As I brooded on all this I felt, as I say, a rising sense of disquiet. Could it be that these stupid officers were right? I recalled the sagacious judgement of Sherlock Holmes. The great detective, you will remember, withheld the facts in the incident of the lighthouse and the trained cormorant because, as he said, it was a case for which the world was not yet fully prepared. Was this also the case with the *Wampanoag?*

Reviewing these matters, I began to feel less of an Elizabethan than an Edwardian, more like Lampedusa than Cellini. I cannot, in fact, quite bring myself to believe the officers were right in their disposition of this lovely vessel, but I think they may have made a point we might find worth considering today. Let me give their line of argument once more. Part of it was couched in the modern idiom of decision making — logical demonstration, mathematical expression, quantifiable elements. They spoke of the ratio of grate surface to midship section and so forth. But then they went on to other considerations. They had the innocence to say they were moved by a *feeling* they had against her; they just didn't like her. And they were simple enough to explain in an official report why they didn't like her. They had a concept of what a seaman was, his obligations, his spirit, his sense of purpose, and this concept was not sustained by men lounging through the watches of a steamer. Furthermore, since their world could be held together by live oak frames and pine planking, there was no need to intrude this extraneous element.

What these officers were saying was that the *Wampanoag* was a destructive energy in their society. Setting the extraordinary force of her engines against the weight of their way of life, they had a sudden insight into the nature of machinery. They perceived that a machine, any machine, if left to itself, tends to establish its own conditions, to create its own environment and draw men into it. Since a machine, any machine, is designed to do only a part of what a whole man can do, it tends to wear down those parts of a man that are not included in the design.

This insight seems to me worth pondering.

Since the launching of the *Wampanoag* we have gone on to refine a system of technical invention that has created a whole new kind of world for us. These inventions have amplified our physical energies, our powers to sense, detect, and communicate, and our powers of thought, often by means of language and other symbol systems, and, more recently, in the computers which expedite and supplant, in certain areas, if they do not actually amplify, our powers to think.

By this system of invention we have multiplied our goods and services and steadily enlarged our advantage over the forbidding physical surroundings. By our developing instrumentation, and the set of ideas that lie behind it, we have slowly moved ourselves from a position of terrifying ignorance and dependence, frail beings in the hands of an angry God, to a place of knowledge and power in which we may think of becoming the measure of all things. This too has its terrifying implications.

For by this system of technical invention we may be on the way to constructing a scheme of things we can no longer live with. In respect to speed, size, complexity, bolts of power, and rates of change, we may soon create a set of conditions that is beyond the scale of ordinary human responses.

If I am correct in all this, our task is to recognize, as the naval officers did, the destructive energy in machinery and to resist it by dealing with it more imaginatively than they did. The task, or one of the tasks, will be, more specifically, to build machinery that will not just do work, but machinery that will continue to work within the field of natural human responses and sympathy.

I can think of places to begin. Professor Feynman at the California Institute of Technology has had the useful idea that we might spend some time scaling down our instrumentation at appropriate places, making things small enough, both in their size and in their effects, to remain in the range of our understanding. We could do the same with speed. Since one of the distorting by-products of technology is the collapse of human time, we could find ways to retard tempos to permit normal responses and accommodations. And also with complexity. Some of our machinery is so complicated we have to spend almost as much time trying to understand it as we would have to spend if we really tried to understand ourselves.

We might also, I think, begin to give attention to the fact that the efficiency in any particular mechanical system is often more than the inefficient human system can stand. Men always have more wandering desires, more irons in the fire at any given moment than the single-minded machine, designed to do only one part of man's work, can ever have. We might begin to arrange things to take this into account. For instance, an obvious place to start is in the central institution of our society, the factory. No laboring man, or almost no laboring man, at a modern machine in a modern plant uses more than a small part of himself in his work; to that extent he endures a subhuman condition which neither high wages, nor shorter hours, nor the displacement of his feeling and energy outside his job can rectify. It is a remarkable fact that a central failure in our industrial society is still, after all these years, the way human beings have to make things in industry. We might therefore think about ways to design and arrange things in our factories so that men could take larger satisfaction, give fuller expression to themselves — even have fun — at making things with machinery, even if they made a few less things, in slightly more time, at a little greater cost in money.

I suppose what I am saying is this. For a long time the design of our technology was determined by our necessity to deal with certain external needs as efficiently as possible, dig more coal, go farther, get there faster, turn out a wider variety of goods for a larger number of people, and things like that, things we required to increase our advantage over nature. The record of achievement here is beyond all reasonable qualification; it is, of course, astounding. But our mechanical triumph may have produced a mechanical atmosphere we can't stand. So we may have reached a point where the design of our technology must take into greater account our interior needs.

That, in a curious way, is what the old naval officers were trying to do. I don't happen to admire their solution, but I respect their awareness that they had a problem. They recognized this problem because they had developed a concept of what they wanted to be, a scheme of obligations, purposes, and feelings, which enabled them to test the machinery put into their midst. We could profit to this extent by their example. But this does not mean that we have to become what C. P. Snow calls Luddites, destroying all the new mechanical possibilities. There are other ways out.

Once I did a good deal of work on the introduction of the pasteurizer. The results of that work are reported in some detail in the last essay in this book. Here it is enough to notice that five quite different societies made use of the instrument and controlled its use in quite different ways. I find this comforting. It suggests that if you have a clear enough scheme of things, a firm enough regulatory system, a culture, you can exert a restraining influence, modify the design of the machinery to keep it doing useful work within the cultural and human boundaries.

I am simple-minded enough to believe we could do the same thing today if we took some thought. But I suggest that recently we've spent a good deal of time on improving the machine and that for a while we ought to concentrate on the other end. The problem is not primarily engineering or scientific in character. It's simply human.

7

"Almost the Greatest Invention"

Part I

A LONG COURSE OF EXPENSIVE EXPERIMENT

In the spring of the year 1862, a rail of Bessemer steel was laid down between two abutting iron rails in the Camden yard of the London and Northwestern Railway.* Four thousand goods cars — about one every twenty seconds — passed over it in the next twenty-four hours. By estimate in the following two years about 20,000,000 wheels ran on the rail. At the end of that time it was still in place and "hardly worn at all." Seven times in the same period the abutting iron rails had been replaced.

Three years later, in May 1865, the first rail manufactured from Bessemer steel in this country was produced at the North Chicago Rolling Mill. In the next two days five

* This essay, somewhat looser in organization and rougher in texture than the others in this volume, was written in 1954. It was to have served as the opening section of a book, never carried further, on the subject of technological innovation.

more were made. Present on the second day were men in the iron trade from six cities, a United States senator, three women, and four strangers. To a man in Wyandotte, Michigan, who had, after three years of labor, produced the steel from which the rails were made, the news fell like "the shadow of a great rock in a weary land." Those in Chicago who attended the actual event believed it of such significance that they arranged to have the first rail "lying in state" for a time in the lobby of the Tremont House.

Those involved had not miscalculated the point of their achievement. In the year 1860 there were in this country five companies that produced between them about 4000 tons of steel. Forty years later, in 1900, this country alone by all known methods produced about 10,000,000 tons. Not all, not even most, of this grand total came from the Bessemer converters, but from this remove it is now possible to see that the single rail in the Camden goods yard and the six rails made by the North Chicago Rolling Mill were the small causes from which flowed such great results.

The nature of those results, imperfectly described by production figures, is perhaps better explained in the words of a man closer to the first events who, in one way or another, observed most of them, was a part of some of them, and appears to have understood almost all of them. In 1892, Abram Hewitt said,

> I look upon the invention of Mr. Bessemer as almost the greatest invention of the ages. I do not mean measured by its chemical or mechanical attributes. I mean by virtue of its great results upon the structure of society and government. It is the great enemy of privilege. It is the great destroyer of monopoly. It will be the great equalizer of wealth. . . . Those who have studied its effects on transportation, the cheapening of food, the lowering of rents, the obliteration of aristocratic privilege . . . will readily comprehend what I mean by calling attention to this view of the subject.

Seventy years after, there may be reservations about the accuracy of this description in detail; no one, however, may seriously question the influence of the large-scale production of steel upon the structure of society and indeed of government. These miracles, technical and social, were set in train when it occurred to a man to blow a blast of air through a bath of molten pig iron.

SOMETHING ABOUT MAKING STEEL

Pig iron is produced when iron ore, coke, and limestone are heated together. It contains, ordinarily, some phosphorous, some manganese, some sulphur, some silicon, some carbon, and a great deal of iron. The same elements but in different proportion are again ordinarily present in steel. The real difference, for our purpose, is that in pig iron carbon is about 3.8% of the whole, while in steel it is about .4% of the whole. The problem for the steelmaker therefore, in very simple terms, is how to reduce easily and quickly the carbon content in pig iron to the smaller carbon content in steel. Until the middle of the nineteenth century, men were solving this problem in about the same ways they had used for literally centuries before. Taking advantage of the fact that carbon has a strong affinity for oxygen, they proceeded by various methods to reduce the carbon content. Only one need concern us here. By packing pig iron in a furnace between layers of charcoal and subjecting the whole to intense heat, a chemical reaction was produced in the course of which the carbon, joining with the charcoal, was extracted from the pig iron leaving what was called wrought or malleable iron or a kind of steel.

This method, like the others used, was slow; not unusually it consumed six weeks. It produced steel in solid, not liquid, form and in severely limited quantities. For these reasons of extended time and restricted quantity, it was expensive. Therefore very little steel, relatively, was produced in the world in the middle of the last century. All Europe, for instance, turned out, by estimate, about 250,000 tons in 1860. At the same time England alone was producing about 4,000,000 tons of the great staple of the metal trade, iron.

The limiting factors in steel production were dramatically disposed of when Henry Bessemer described, in 1856, his own new process. He reported that if a blast of air was sent through a mass of molten pig iron, the result was not a cooling of the mixture, but a violent combustion by which the carbon and some of the minor elements in the pig iron were removed. What happened was this: the silicon, manganese, and carbon joined with the oxygen in the air to form combustible materials or gases that burned away to leave a relatively pure iron behind.

There is only one other thing that needs notice here. Commercial steel needs some carbon and manganese in it. Bessemer tried, at first, to leave sufficient quantities of these elements in the final product by stopping the blast of air before combustion was complete. Using this rather haphazard procedure, he could not control his quantities within acceptable limits. Some time passed before someone else hit upon the happy idea of burning out all the carbon and manganese before adding back the elements in desired and precise quantities. This addition, in the form of a triple compound of carbon, manganese, and iron, is called the recarburizer, or spiegeleisen. The commercial success of the Bessemer process depended directly upon the development of this idea of controlled recarburization. The significance of the total process thus described lies in this: that by its use it was possible for the first time to produce large (two and one half tons) quantities of liquid steel that could be poured into any desired shape in the space of twenty minutes.

Before proceeding it should be pointed out that the foregoing description, while adequate for purposes of explanation and accurate as a description of the general theory involved, is by no means precise or complete. For instance, there are far greater variations in the composition of iron ore, pig iron, and steel than have been suggested here. There has been no real attempt to explain in detail what really takes place in a Bessemer converter in the course of a "blow." A whole train of intricate and interesting chemical reactions occur during this exciting twenty minutes. Some idea of what has been left out and some idea, too, of the elegance of the necessary procedures can be gained from the following statement concerning one of the more obscure reactions that is involved in the conversion of iron to steel. "The behaviour of manganese and silicon," said Alexander Holley, "as well as their proportion in the process of steel making was brought by a long course of expensive experiments from the realm of speculative guess work to that of an exact science."

HOW AN IDEA GETS AROUND

With something of the process and something also of the significance of the process in mind it is now possible to investigate the origins of the idea. One day in 1855 Henry Bessemer, while packing a furnace with pig iron which he was going to melt down, inadvertently put two pigs in a place where they were exposed to the blast of air he was using to produce a draft instead of to the direct heat of the fire. On opening the furnace door after a suitable interval, he noticed for the first time these two pigs, and he noticed also that they had become "thin shells of decarburized iron" — that is, Bessemer steel. Confronted by what he himself called "the remarkable incident," he went on to his great conclusion that "atmospheric air alone was capable of holding decarburizing gray pig iron and converting it into malleable iron [Bessemer steel] without puddling or any other manipulation."

Every Darwin, apparently, has his Wallace. Eight years earlier, in 1847, another man had reached a similar conclusion by similar means. William Kelly, a maker of kettles at Eddyville, Kentucky, had packed the "finery fire" one day in such a way that a pig "at the edge" was exposed more to the air than to the heat. When he opened his furnace door, he looked at the unmelted metal — decarburized iron — and he too went on to the generalization, as he said, that "air was fuel."

This seems a convenient moment for some further reflections on Pasteur's notion that fortune favors prepared minds. First a word about fortune itself. In this case it had been at work for some time; that is, "steel was sometimes made as an accidental product" in a good many old forges and furnaces long before Bessemer and Kelly peered through their furnace doors. Inadvertences, such as those described, had produced it, but the minds observing the accidental product were not prepared to reach the conclusions arrived at by Bessemer and Kelly. It may be useful to investigate, briefly, the special qualifications they possessed. There were to begin with their own personalities and cast of mind. As anyone who cares to read the books about them will quickly discover, they were picturesque characters, inventive types. Bessemer was the more successful and had the longer record in the field of invention, but Kelly, before deciding "air was fuel," had designed his own unique procedure for making his iron kettles. These are superficial similarities; they shared more substantial ones. Both, at the time they made their generalizations, had begun to ask themselves questions about the ordinary procedures they had been using in their work. Bessemer had been requested by the Emperor of France to seek means to improve the metal in French ordnance. Kelly had no such specific intent, but he was much troubled by the fact that the wood he was using to make his charcoal was about exhausted. It is too much to suggest that he was looking for a substitute, but it is quite obvious that the absence of charcoal in the immediate vicinity presented him with a problem he had to solve. In both cases, therefore, it is fair to say there was some preoccupation with new solutions for an old problem. Finally, and perhaps most important of the personal similarities, neither man had grown up in the iron trade. Bessemer came to his experiments for Napoleon III at forty-two after attending to such matters as the design of machines to make glass, velvet, sugar, and bronze powder. Kelly was in his thirties when he drifted out of dry goods and into the kettle business. They were in a sense, to borrow a phrase from the sociologists, marginal men — that is, men on the margins of an art or profession and hence without the commitments of those with long training or experience to the rituals of the calling. To put it more positively, they were possibly freer to discover meaning in accidental deviations from normal results — remarkable incidents — simply because their thoughts and expectations had not been steadily confined within the boundaries of common practice.

It was not only these special personal qualifications that enabled these men to take such impressive advantage of the opportunities fortune placed in their way. They inhabited a favorable environment. To begin with, the state of the art, after generations of stability, was beginning slowly to change. In the immediately preceding years blast systems had been improved, bigger furnaces built; there was a growing understanding of the place of chemistry in metallurgy; with the gently rising tide of scientific knowledge more systematic procedures for observation and investigation had been devised. In other words, an environment of appropriate information and instrumentation was being assembled in which there was an increasing opportunity for someone not only to hit upon the idea that air was a decarburizing agent but to pursue and develop it before a favorably disposed audience. Indeed in the years from 1845 to 1856 there were other men in England, the United States, and Germany who were, rather more consciously than either Bessemer or Kelly, experimenting with the use of air in the making of steel. They failed, however, and it is interesting to speculate on whether this failure occurred primarily because they were proceeding more consciously and systematically to reach the grand theoretical explanation provided by both Kelly and Bessemer.

A second favorable element was in the environment—an accelerating demand for steel. New means of transport, like railroads; new methods of communication, like the telegraph; and new possibilities in construction, as in bridges and buildings, were creating new requirements that iron could not wholly satisfy. Thus there was presumably an audience ready to listen with more than intellectual enthusiasm to the findings of the two inventors.

There is a pleasing footnote to these reflections on the interaction among fortune, prepared minds, and favorable environments. Attention has been given to the things Bessemer and Kelly could do under appropriate conditions. It is of equal interest to discover what they could not do. Earlier it was said that the commercial success—that is also the technological success—of the process rested upon the addition of spiegeleisen to the decarburized iron. Both men had months of discouragement, after reaching their initial conclusion, in trying to control within acceptable limits the amount of carbon in their final product. They attempted to figure out ways and means to stop the reactions in the converter in time to leave a correct quantity of carbon uncombusted. Since they didn't know enough about what was going on in the converter, they failed. Now the solution to the problem is, of course, of the utmost simplicity, of a kind that used to win for the lady from Philadelphia the open-mouthed admiration of all the little Peterkins. Run the blow until all the carbon is extracted from the pig iron, and then add back the desired amount. Neither Kelly nor Bessemer thought of this. It occurred to a man named Robert Mushet, who first poured the triple compound of carbon, manganese, and iron into the converter after a blow. And here too there is the intercession of a remarkable incident. For some years Mushet had been experimenting with spiegeleisen, but until the Bessemer process was announced he had never figured out anything very useful to do with it. History, as T. S. Eliot would say, has many cunning passages.

The way is now clear to discover how knowledge of the new process for making steel entered the public domain. Kelly, it will be recalled, decided that air was fuel in 1847. During the next ten years he built a series of small experimental converters near his factory in which to test his conclusions. These converters were constructed in a small copse supposedly to hide their existence from Kelly's father-in-law, who believed the inventor should concentrate on the making of kettles. Actually, an ingenuity even greater than Kelly's would be required to conceal a converter in full blow from a suspicious bystander, and as a matter of fact, word of Kelly's activities of course went abroad. So, in time, did some of his product in the form of small billets and boiler plate. The response to the spreading rumors and this tangible evidence was not especially encouraging. Though men as far north and east as Cincinnati became aware that Kelly was at work in the backwoods of Kentucky, nobody except his kettle customers expressed a direct interest, and they asked him to return to the old and approved ways of making kettles. In 1857 the panic wiped out the business and terminated the Eddyville demonstrations. Kelly himself departed for Pennsylvania, where he took a job at Johnstown with the Cambria Iron Works. There, in the mill yard, he gave two further demonstrations of his procedure. The first was disastrous: the air blast blew the molten metal high up in the air, and the resulting display, called Kelly's fireworks, provided merry recollections in Johnstown for years. On the second attempt he did succeed in producing several small globules of steel and a slightly larger steel plate. These results did impress the general manager at Cambria, a man named Daniel J. Morrell, but since the ironworks was then fighting for its existence amid the perils of 1857, he did nothing for the moment.

The results of a decade of demonstration were these: some men in the border states had heard by word of mouth that a man in the Cumberland valley was trying to make steel by using an "air boiling" process; some customers for this man's kettles had asked him to cease his experiments; some curious bystanders at Eddyville or in the mill yard at Johnstown had been momentarily entertained, terrified, confused, or impressed. After ten years, Kelly, on the available record, had reached the intelligent attention of one man in the iron trade who might, had circumstances been otherwise, have done something about the new process.

In England the thing was better done. Bessemer like Kelly continued after his initial discovery to work at his idea. Like Kelly he built an experimental converter from which he soon produced acceptable steel in small quantity. In the summer of 1856 he showed a piece of this steel to the distinguished engineer George Rennie. It chanced that in that year Rennie was chairman of the mechanical section of the British Association for the Advancement of Science. He therefore persuaded the very reluctant Bessemer to present a paper describing his findings before the professional society. Accordingly, on August 11, 1856, at the meeting of Section G of the British Association Bessemer gave his systematic analysis, "on the manufacture of malleable iron and steel without fuel."

This title and the concept had produced a certain amount of innocent laughter in the lobby before the meeting of Section G. Following the talk, however, the audience remained to pray. "Gentlemen," remarked the great James Nasmyth as he fondled a small piece of steel Bessemer had used in demonstration, "this is a true British nugget; its commercial importance is beyond belief." The extent of the subsequent publicity was also hard to believe. British newspapers and trade journals gave the paper full and accurate report. So indeed did the newspapers and journals on the Continent. And five weeks after Bessemer spoke at Cheltenham the New York *Tribune,* September 16, 1856, devoted three notices within a week to a complete description of the new process. Two months after the original presentation, therefore, the news had traveled through the communication circuits of the iron trade and reached the general public. Very shortly, action was taken on the information thus supplied. Converters were built in England and France in 1856, in Sweden in 1858, in Germany in 1862, and one even at Madras, India, in 1861.

The contrasting experiences of Kelly and Bessemer raise for reflection some nice points about the vexing relationship between theory and practice and about communication. For instance, it seems quite clear that many of Kelly's observers were put off by the ineptitudes of his demonstration — the discontinuities of performance, the fireworks, and so forth. Not only were they put off by all the side effects attending any experiment; they were not, by the demonstration, informed of what was actually going forward. The void separating Kelly's audience from Bessemer's is well described by the chance remark tossed off by Rennie when he was seeking to persuade Bessemer to give his paper in spite of the inventor's feeling that the thing was not yet a predictable commercial success. "You ought not," he said, "to keep it a secret another day. All the little details requisite will come naturally to the ironmaster." As Bessemer was to learn to his pain, the adjustment of little details came with small natural grace or ease to the ironmasters in the succeeding years. But Rennie had grasped that technical considerations were, in relation to the concept, but minor matters. The essence, the great advance, the remarkable opportunity lay in the soundness of the idea itself. Furthermore, the episode suggests that as a means of communicating new ideas it is more useful to rely on the description of a general theory that makes sense than it is to depend primarily on practical demonstrations. The danger in the latter is that practical demonstrations may leave the observer with the feeling that at best the demonstrator can do something that the observer himself cannot, while the advantage of the former is that by placing the theoretical justification before the observer it may enable him to do something he could not do before. All this of course presupposes a prepared audience to which one can naturally turn. In England there was the British Association and beyond it a fairly stable iron trade that was one of the main supports of a fairly well developed industrial economy. In this country there was no similar association and only a struggling iron trade that was pushing itself slowly forward into an economy that was in a far earlier stage of industrial development.

The foregoing pages have been concerned with the origin and revelation of a new procedure. Actually they have dealt with what may be called "the learning process" — how and under what conditions a new idea is received in the mind of the innovator and by what means and in what circumstances it is translated to a wider audience. Stated still more briefly, the subject is how people find out about things. The evidence in this case may be unfamiliar and some of the observations unexpected, but it can hardly be claimed that any of the reflections are particularly new. An editor of the *Tribune* in 1856 appears himself to have known it all. After the complete and careful descriptions of the Bessemer process appeared in his paper, a flood of letters descended upon his office dedicated to the proposition that the thing was either a trick or a lie. Confronted by this evidence he concluded that "every innovation is looked upon with suspicion, the more so as it is important in its results but simple in its nature. This appears to be the case with the invention of Mr. Bessemer — an invention or discovery toward which others have approached so closely that it is difficult to understand how they failed to arrive at the result." And why, he ended, has our iron trade stood still? "Why should not our large iron masters have laboratories attached to their works, as is the case abroad. . . . Until our manufactories are scientifically as well as practically studied our progress must be slow, as accident alone in one case leads to improvement and deductive reasoning in the other."

TRYING TO CONTROL THE EXPERIMENT

Thus far the concern has been primarily with aspects of the intellectual development of the Bessemer process. There is now the question of its commercial development. Examination of this question will be confined here to how men sought, in the first stages, to apply the process in the production of steel in this country. This period may be called the experimental period. The first steps in this survey must be taken into the United States Patent Office. Thither in the fall of 1856 went a representative of Henry Bessemer with a claim for "the conversion of molten crude iron or remelted pig . . . into steel . . . by forcing into and among the particles of a mass of molten iron currents of air or gaseous matter." In support of this claim there was attached a sketch of the machinery used. This converter was a large vessel shaped like a pear with small holes, called tuyeres, in the bottom. It was mounted on trunnions. To make Bessemer steel the converter was filled with molten pig and air was blown upward through the tuyeres thus setting off the violent chemical reactions by which the carbon in the pig was extracted. When the reaction was over, the converter was tipped on its trunnions to pour the metal from the mouth at the top of the pear-shaped vessel into molds.

On November 11, 1856, a patent was granted for this claim to Henry Bessemer. Then great things began to happen. An attorney for William Kelly appeared in the office of the commissioner within a few days to "pray that an interference be declared [against Bessemer] and that an opportunity be given me to prove priority of invention."

The opportunity was given. On April 1, 1857, over mark and signature a cloud of witnesses deposed that they, as workmen, had helped Kelly make steel with air in Eddyville, Kentucky, as early as 1847, "the year the Cumberland overflowed its banks." Such evidence convinced the commissioners, who on April 14 declared that "priority of invention in this case is awarded to said Kelly." Accordingly, on June 23, 1857, a patent was granted to him on his claim of "blowing blasts of air, either hot or cold, through a mass of liquid iron." With this claim the inventor had presented a sketch of a primitive converting vessel that did not tip on trunnions, but remained upright and stationary. Just before this Kelly patent was issued, Robert Mushet obtained, on May 26, 1857, an American patent for the use of spiegeleisen.

There is a temptation to linger a little longer in the Patent Office amid its entertaining documents. For instance, there is a letter from Charles Cargill urging the commissioner to reject Kelly's claim on the ground that he himself had the principle "fully matured as early as 1842." Three days after writing the letter, however, on April 1, Cargill appeared before the commissioner in behalf of Kelly's claim for a patent on the process. There is also a communication from Bessemer's lawyer rebuking the commissioner for "that unjust and obnoxious decision which prevents Mr. Bessemer from going behind his *foreign* patent or a printed publication in order to prove the date of his invention." And there is also a commissioner's ruling whereby it was decided that J. G. Martin had first thought of blowing air through molten metal contained in a tube, while Kelly had first thought of blowing air through metal held in a ladle.

These documents and other pieces of germane literature

do suggest something about the ultimate refinements of assigning, by patent, credit where credit is due; but they are, here, of only incidental interest. It is necessary only to know that by the middle of 1857 the several parts of the process — that is, the concept of the use of air, the machinery, and the indispensable additive — were all covered by American patents. But the patents were separately held for, it should be noticed, while Kelly had been granted an interference, the Bessemer patent, with its identical concept and superior machinery, remained on the books. There was thus, as a result of this nine months' activity in the Patent Office, a certain ambiguity about what rights one really needed in this country to begin the manufacture of Bessemer steel.

How the rights were obtained, how and where the first Bessemer steel was manufactured in this country, and how the patent ambiguities were finally resolved, these are the principal elements in the period of early experiment. Some of these elements, especially those having to do with the patent situation, are, in themselves, really quite dull and dreary. Any account of them must contain a series of names, dates, and details that in the absence of any interesting larger context are hard to hold in mind. Fortunately, the account, virtually interminable if fully developed, can be reduced to a more assimilable brevity.

It will help in keeping matters straight during the next few pages to think of the Kelly patent as covering the process, the Bessemer patent as covering the machinery, and the Mushet patent as covering the indispensable additive. There are possible refinements and qualifications to this broad statement, as there always are in patent cases, but this is the heart of the matter. It will help, too, to bear in mind that anyone wishing to make Bessemer steel had to possess all three patents.

The first date of importance is 1862. In that year a group of men — E. B. Ward, Zoheth Durfee, Daniel J. Morrell — set out to buy both the Kelly and Bessemer patents. In the same year these men hired W. F. Durfee to begin the construction of a Bessemer plant at Wyandotte, Michigan, about ten miles from Detroit on the grounds of a blast furnace and rolling mill owned by E. B. Ward. When W. F. Durfee began work he had never seen a converter, nor had his sponsors obtained any patent rights. But since Zoheth Durfee, William's cousin, was in England and confident of obtaining Bessemer's patents, it appeared only "to be anticipating the acquisition of property rights . . . to use such of his [Bessemer's] inventions as seemed suited to the purposes in view." Accordingly, W. F. Durfee set to work with a copy of Fairbairn's *History of the Manufacture of Iron,* drawings of Bessemer's machinery, and his own sense of the "internal fitness of things."

While Durfee was slowly erecting his plant in 1862 and 1863 the men who had hired him were actively in search of patent rights. Those of Bessemer were refused to them at the end of 1862, but after interminable negotiations with members of the Kelly family they finally, in May 1863, obtained rights to the Kelly patent. Then through the skillful diplomacy of Zoheth Durfee, they also obtained the American rights for the Mushet patent. On the foundation of these two patents, the Kelly Pneumatic Process Company was organized. By the time its organization was complete, as a copartnership, its membership included, besides Ward, Zoheth Durfee, and Morrell, Messrs. Chouteau, Valle, and Harrison, of Missouri, who owned the ore bed from which the iron came, James Park of Pittsburgh, a Detroit friend of Ward named Lyon, Robert Mushet and two of his friends, and the four daughters of William Kelly. The passengers in the omnibus obviously increased every time someone who held in his hand some part of the process — whether a patent right, a blast furnace, or an ore bed — showed up with friends. Nevertheless, E. B. Ward and D. J. Morrell remained the determining influences.

Under their supervision W. F. Durfee in the late summer of 1864 finished the plant "to test by an experiment on a large scale" the pneumatic process for the production of steel on a commercial basis. He had put together in two and one half years, and largely out of his own head and his sense of internal fitness, the blowing engine, the casting ladle, the hearth for melting the pig iron and spiegeleisen, the cranes to move the metal in both fluid and solid form, and the converter itself. This was a considerable achievement, and to it must be added the fact that as part of the plant Durfee built the first real laboratory connected with an iron- or steelworks in this country.

With this equipment "in one of the early days of September 1864" the Wyandotte works produced the first Bessemer steel on a large scale (two tons) in this country. It should be pointed out here that this steel was made in a converter that clearly invaded the Bessemer patent rights and that it was produced six weeks before the rights to Mushet's patent had been obtained.

Now while all this was going on in Michigan, an experiment, virtually parallel, was being conducted in Troy, New York. In 1863 a man named Alexander Holley had persuaded Erastus Corning, John A. Griswold, and John F. Winslow, partners in several large undertakings, to investigate the Bessemer process. For them he, like Zoheth Durfee, journeyed to England, returning in the early months of 1864 not with the rights to the Bessemer patents but with a license to manufacture Bessemer steel on a royalty basis. Also, by the agreement signed December 31,

1863, he had obtained a three-year option to buy the Bessemer patent rights in this country. On this foundation a partnership, never defined on paper, of Griswold, Winslow, and Holley began to erect an experimental steelworks on the grounds of, but separate from, the Albany & Rensselaer Iron Works, in which Griswold and Winslow had a large interest.

Here, in Troy, New York, Alexander Holley, by taking thought and with infinite pains, constructed a works not dissimilar in character to that in Michigan, save that the first blowing machine ran by water power and not by steam. On February 15, 1865, five months after the great event at Wyandotte, the Troy works made their first blow. And here too it may be noticed that the steel was manufactured only by violating the patent rights of Robert Mushet, now held in this country at Wyandotte.

So, early in 1865, ten years after Bessemer had stumbled upon his discovery, both experimental works in this country had demonstrated that steel could be made in large quantity, and each had succeeded in doing so only by violating the patent rights held by the other.

How long, in the absence of outside stimulation, this situation might have continued it is impossible to say. The men at Wyandotte had tried, and continued to try through 1865, to design a successful converter that would not have to include the "tumbling" quality which was protected by the Bessemer patent. They failed. On their part the men at Troy tried equally hard to develop a substitute for spiegeleisen, but all experiments, principally with a chromium derivative, failed. Neither side, in this "state of affairs [which] caused great jealousy and rivalry" dared advance upon the other or, with success so near, desired to capitulate. Happily, outside stimulation appeared. A "group of Pennsylvania people," impressed by the actual production of Bessemer steel in two different places, approached the Kelly Pneumatic Process Company to obtain a license to manufacture steel on a royalty basis under the Kelly patent. The legal counsel for these people then examined the general patent situation and decided that what his clients needed most was a license under the Bessemer rights. He took the matter to the lawyers for the Troy people, the firm of Blatchford, Seward, and Griswold of New York. This firm then first examined the possibility of bringing suit against the Wyandotte company. It was then decided that a suit by a company that was itself infringing the Mushet patent would not resolve the problem. And yet something had to be done; steel could be made, and there were people anxious to make it.

Recourse was had to *Force majeure*. In July 1865 Bessemer took out, for the first time since his original patent, five new American patents on machinery, duplicates of British patents taken out from 1860 to 1865. Then Winslow and Griswold, through their counsel, Blatchford, took up their option of all Bessemer's patents for a sum of £8000 on December 7, 1865. In the first week of January 1866 they then granted a license under these patents to "the Pennsylvania group." Thus buttressed with patent rights, a plant in being, and a license to another interest (albeit this license did not include the rights to spiegeleisen), Winslow, Griswold, and Holley sought an agreement with the men at Wyandotte.

The latter were in no cheerful mood. W. F. Durfee had left early in 1865 after his laboratory had been dismantled by what he called "a syndicate of sin." His place as manager had been taken by three different men in rapid succession. None of them had made much progress in the design of a new converter. Beyond that disagreements had broken out among the large number of copartners; some wished to expand, some wished to contract. Confronted with the strength, organizational, financial, practical, and legal, of the Troy group, the uncertain structure of the Kelly Pneumatic Process Company disintegrated. An agreement was reached whereby the three patents — Kelly, Bessemer, Mushet — were pooled. The trustees who were to administer the licensing of these patent rights were Winslow and Griswold of Troy and Morrell of the Wyandotte company. All profits were to be split 70% for the Troy people and 30% for distribution by Morrell to the various Kelly interests. When this agreement was signed, after six months of debate and negotiation, in the fall of 1866 it was looked upon, as it has been ever since, as a great mystery. How could the Wyandotte people sacrifice two thirds of the patents for one third of the resulting profits? The mystery is not apparently very hard to penetrate. The jerry-built structure of the organization, the technical difficulties, and the divided interests reduced the will and potential of Ward and most of his associates. One of these associates however got what he wanted. Ever since Kelly had produced his fireworks in the mill yard at Cambria, Daniel J. Morrell had wanted to make Bessemer steel. He had constructed a little converter in 1861; he had joined the Kelly Pneumatic Company in 1862; and now he was one of the three who put all those pieces together that were needed to make steel, and who held these pieces in their hands. As Blatchford wrote off to Winslow in 1866 thanking him for a cask of whisky sent at the conclusion of the agreement: "now after all this work, perhaps you can make some money." With this letter the experimental period came to an end. By December 1866, steel, in quantities suitable for commercial purposes, had been produced by two companies; the rights to manufacture had been consolidated; a third company had been licensed to begin operations. The period of expansion — more precisely, of development — was at hand as a new year opened.

Before turning to this new period, it will be well to examine more closely some of the events of this experimental time; to see what further material may be laid on the simple scaffolding of dates and names that has thus far been constructed.

It will be satisfying to get the patent question out of the way at the outset. Someone with imagination and great stamina should address himself to a study of the place of patents, historical and desirable, in the innovating process. No such study now exists. There are lots of opinions, as for instance this one offered by a sometime United States Commissioner of Patents to the National Association of Manufacturers in 1951. The contention was that "the American patent provides a condition under which people are free to make whatever contribution may be within their capacities, and has produced the highest degree of liberty obtainable in the development of our citizens." These are large, strong, and, in the instance under review, inappropriate words. There can be no doubt that in this case the patent system extended the period of experiment, complicated the task of the experimenter, and signally failed to protect the rights of the man to whom the Commissioner of Patents assigned priority of invention. Indeed, during this period, when the exchange of data was of the greatest importance, the patent system transformed the event into something like a military occasion in which pieces of highly classified information were withheld from enemy forces. The situation was salvaged by the kind of irony so frequently discovered in military history. Breaches of security reduced the whole system from a shield of protection to an expensive nuisance. A lot of useful energy at both Troy and Wyandotte went into an attempt to find substitute solutions for converter design or spiegeleisen that might better have been concentrated on improving known designs and methods. Actually, the urge to make steel proved so great that both sides achieved the necessary degree of "liberty" and freedom by disregarding the inhibitions of the patent system. It was only when steel had been made that men began to have a high regard for rights conferred by patents, and then that regard was used to establish a stable corporate structure for the control of future production.

Perhaps the most striking single thing in this whole experimental period is the fact that so few men came forward to explore the new possibility. It would be natural to expect that those in the iron trade would reach out eagerly for the process. Here was something understood in 1855, at the very moment of its creation, to be the source of "a perfect revolution . . . in every iron making district in the world." Here was something recognized in 1856 as possessing "commercial importance beyond belief." Here was a procedure that by 1857 had been fully described and transmitted around the world. And yet, by 1863, seven years after the Cheltenham paper, only four men or groups of men in this country had seriously investigated its possibilities. Since anyone in those years could have obtained licenses, as opposed to outright purchase of patent rights, this general reluctance is even harder to understand.

The first explanation that will occur to mind is that the risks involved were so great that they deterred all but the most valiant. There is, to be sure, some evidence bearing on this point. In the eighties and nineties witnesses before tariff commissions or congressional committees were wont to enlarge upon the terrors of these times. Like Desdemona they would admire the pioneers for the dangers they had passed. They talked largely of millions, described the "total loss" of early plants, and mentioned "huge sums" spent in litigation. The effect was of whole fortunes placed at the hazard.

Testimony of this sort no doubt served a useful purpose at the time, but, as a matter of fact, it was not at all accurate. Before going to England in 1863 Holley wrote a friend that if he were successful in obtaining the Bessemer license, Winslow and Griswold expected "to spend about $38,000 in establishing the manufacture." It does not seem likely that Holley was referring to the cost of patent rights, and, in any case, his figure is quite close to the contemporary cost for the construction of small plants comparable in size with the two experimental Bessemer works. Expenses, either at Troy or Wyandotte, could hardly have exceeded, counting everything, $60,000 or $70,000. And parenthetically it may be added that the only expenses for litigation, besides some travel money and a cask of whisky, appear to be Blatchford's bill to Winslow. His legal fees from February 1, 1865, to the fall of 1866 amounted to $2594.06.

So the risk in financial terms seems, even for those days, very small. For instance, six men in Pennsylvania in 1854 put up among them $180,000 to reorganize an ironworks that had failed before it gave any promise of real success; and in 1862 Winslow and Griswold had promised Abraham Lincoln that they would assume all costs of the ironclads if the ships were failures. These costs were about a quarter of a million dollars.

One must look further for some satisfying explanation. To say that the risks were not financial is not to say that there were no risks. It was predicted, as already noticed, in 1855 that with the coming of the converter "a perfect revolution threatened in every iron making district in the world." Few men look forward cheerfully to even a perfect revolution, and there were excellent reasons why the men in the iron trade in the fifties and sixties should look with disfavor on any dramatic change in their fortunes. In those years they were laboriously assembling the several elements of what they hoped would be the modern iron industry. They had been at work taking the many widely separated blast furnaces, forges, bloomeries, and rolling mills out of the backwoods where they served the farmers' needs and putting them together in or near the cities. In so doing they were conducting, in some sort, a revolution of their own. True, this revolution had been largely forced upon them by the rising demand for iron throughout the country, but they had acted in the new conditions with skill, wisdom, foresight, and some imagination. By the Civil War great progress had been made toward the objectives of centralization and increased production. At Pittsburgh, for instance, there had been fourteen mills consuming 54,000 tons of pig iron in 1850; seven years later twenty-one mills used 132,000 tons of iron.

The center of all this creative activity was Pennsylvania. There, then as now, one may seek the dominant mentality in the industry. And, for the years under review, the prevailing attitudes and concerns may be discovered most easily at Johnstown. What happened at the Cambria Iron Works in that city can be taken as a description of what was happening throughout the whole state. Cambria was started in 1852. For six hard years, through fire, financial panic, two reorganizations, and continuing technological difficulties, the men at Cambria labored to build an organization on sound financial and mechanical foundations. By 1858 they had accomplished their not inconsiderable purpose. Before them stood an integrated ironworks of high capacity — blast furnace, bloomery, rolling mill, and the great three-high rail mill of George Fritz. This last was an impressive innovation, a device that made two rails grow where one had grown before. In those six years the owners had demonstrated all the admirable qualities: faith, fortitude, energy, resilience, and imagination. They had likewise demonstrated their distaste for the Bessemer converter. From 1857 when it was first brought to their attention by D. J. Morrell, the general manager of the works, right on down to the close of our experimental period, the men in power at Cambria resisted all efforts to interest, persuade, or cajole them into an investigation of the new process.

The resistance of these men can be, and has been, explained in several ways. Concentrating, as they were, all their energies in an attempt to place a precarious venture on a sound basis, they had no excess time or energy for irrelevant considerations. They were conservative, and the thing was new. They were too blind to see the opportunities or too dull to understand them. They were men in the iron trade, and one of the first rules taught by Tubal Cain was that the trade stands still. Many of these explanations are mere descriptions; none of them is very satisfying; and few of them are even very true. In seeking a more convincing explanation, it may be useful to begin with the assumption that the men at Cambria did actually grasp the meaning of the new idea but were unready to run the risks involved in accepting it. They were, it has been pointed out, already engaged in a kind of revolution of their own in which they had revealed splendid qualities of courage and resourcefulness. But in that revolution they were working always with familiar elements. They were simply taking well-known objects like blast furnaces, bloomeries, rolling mills, modifying and improving them separately or putting them all together in a new system to produce better the old things. They worked always, therefore, with well-known quantities within a familiar environment; at each step they could calculate and control the effects of their decisions and actions. The converter, however, introduced a new element with strange characteristics and undescribed energies. Its operation could disturb, indeed change, the environment in ways they could not predict, understand, and therefore control.

And they were asked to embrace the strange new procedure as a free act of the will. The integration of the iron business they had undertaken in response to the rising clamor for iron. No similar demand, insistent and immediate, for steel beset them. To accept the converter meant that voluntarily they would yield up their rights in a known and calculable and rewarding market to enter a market with requirements unknowable for uncertain rewards.

Finally they were caught up in and dependent upon an intricate net of relationships with employees, competitors, and customers, relationships slowly and carefully constructed around the manufacture and sale of iron. By hard work and infinite pains, they had reached a position of eminence in an ageless guildlike craft and in the community. Now they were suddenly confronted not by a new way to make iron, but by a new way to make something that would replace iron. Would not this new thing destroy the competitive advantage, so hardly won, by forcing every man to start over again in the same degree of innocence and from the same level? Would it not, by replacement of an old reagent, iron, with the new element of steel, replace also the customs, habits, procedures, and hierarchical arrangements upon which the security of life in the iron trade depended? The converter, in this context, looks less like a tool of commerce and more like some catapult leveled against a walled town.

It is difficult to explain in any other satisfying way why so many men who stood, as is now apparent, within easy reach of the cornucopia in 1857 failed to lay their hands upon it. The explanation becomes even more persuasive when one turns to examine the few men who did attempt to acquire and develop the Bessemer process in this country. The important names to remember are these: at Wyandotte, E. B. Ward, D. J. Morrell, James Harrison, C. P. Chouteau, James Park, and the two Durfees; at Troy, Erastus Corning, John Augustus Griswold, John Flack Winslow, and Alexander Holley. Of these the most important by far, as promoters, were Ward and Winslow. W. F. Durfee and Holley provided the engineering intelligence and will be treated separately later on, but much of what is said here applies also to them. It is unfortunate that time cannot be taken to investigate each of these individuals in detail. For one thing, there are always dangers in collective characterization, and, for another, these men, taken separately, offer both fascinating and rewarding subjects for study. But for the purposes in view, the generalized investigation must serve.

It will be useful to begin by citing the fact that the two experimental plants, Wyandotte and Troy, were astride, really on the fringes of, the great iron-producing district of the country. This peripheral position is at least symbolic. Though it is too much to say that all the men concerned were on the edges of the iron trade, it is accurate to suggest that the trade represented in virtually every case only a part of the total interest of the man involved. Among the other pursuits that engaged them were mines, shipping lines, a newspaper, timber, real estate, livestock, grain, fur trading, politics, and railroads. As Cambria summed up the state of the art in Pennsylvania, so E. B. Ward in his operations suggests the manifold activities of the group.

Starting from nothing about the year 1840 he put together a kind of industrial margravate that included seventeen ships in the Great Lakes, four rolling mills in the Middle West, a principal interest in a paper of which Carl Schurz was editor, 80,000 acres of Michigan timber, 2000 acres of mining land in Wisconsin, a large tract of Detroit real estate, and the Pere Marquette Railroad, of which he was president. Ward is, to be sure, an exaggerated symbol of diversity and success, but he is not inaccurately suggestive. These men, if they were not in everything, were at least in a great deal. They were the product of the dazzling explosions in our economy in the fifties and sixties by which the great energies and resources and possibilities were opening up for the first time. Among the multiplying opportunities it was difficult to select; no man knew what causes would lead to what effects. In such conditions it occurred to some bold men that good sense required them to play the field without trying to select the favorite. And so they were led through a system of logical connections — mines led to furnaces, and furnaces to rolling mills, and mills to railroads, and railroads to real estate, and real estate to timber. There wasn't time and there wasn't the desire to concentrate on the single thing.

Experience like this leads naturally if not to a restless at least to a fresh and inquiring spirit. John Flack Winslow, for instance, began as a clerk in a commission house, entered the iron business in New York, joined with Erastus Corning, first president of the New York Central, became an expert on railroad construction, turned during the war to the design and building of the *Monitor*, served after the war as president of the Rensselaer Polytechnic Institute, became president of a small railroad, and then of a bridge company.

All this is not to suggest that these men were propelled

along and upward by the force of a series of accidental explosions. There was a natural logic in the selection of their ventures and a careful calculation, ordinarily, of the chances run. The Bessemer converter is a case in point. Running through all these biographies is the theme of the railroads — Holley had written a book about railroad practice, Corning, Ward, Winslow, D. J. Morrell were railroad presidents, James Harrison negotiated the $70,000,000 loan which bought the Pacific Railroad from the state of Missouri, and so forth. The railroads were not only one of the greatest customers for iron and steel, but they were also searching in the fifties and sixties for means to improve the iron rail. When the converter suggested a solution to the problem, it was natural for men familiar with railroads, and also with the production of iron rails, to move out toward the new process.

It is a temptation to remain longer in the company of these men. One cannot read their letters, or the recollections of those who knew them, or their biographies in local histories or newspapers, or by all the other usual means follow their paths across this period without beginning to feel what an attractive and interesting group of men they were. No doubt they saw the main chance, but they also had their eye on the little pleasures, and they could express themselves and their ideas in clear and graceful language, shot through with scraps from the Old Testament, the English poets, and, for "the language of Troy is always classical," lyricists even more remote. But more important, they appeared to have not only some buoyant sense of the tremendous opportunities in their environment but a sustaining image of themselves as members of a community — as mayors, college presidents, organizers of scientific expeditions, and the like. It was not so much their money (and they were for the most part very rich men) that they spent and gave, as their energies and judgement. And these they employed constantly in widely dispersed areas. Taken all together there is a kind of warmth and dignity in their careers. It has been a pleasure to come upon these rounded, public-spirited, resilient, and cheerful men amid the legendary shades of the narrow, grim, compulsive types that are believed to have succeeded them. No doubt it was the times.

These were the innovators, the men without heavy commitments to the system of attitudes and prejudice built up in the iron trade. By virtue of the diversity of their interests they were, relatively, freer to assess the meaning of any new process and to throw their energy and intelligence behind its development. They could weigh the merits of Bessemer steel not by trying to decide what it would do to them in the iron trade but what problems it would solve for them in their railroad interests. It is interesting to discover the consistency that develops in any personality or attitude. The history of the Bessemer converter is a case in point. An almost careless quality, a preoccupied note, runs through the activities of these men as they strove in the sixties to put the innovation on a sound basis. This quality is maintained until the end. After the experimental period closed in 1866 there seems to have been a loss of interest on the part of the prime movers. Ward ceased to have an active part and turned away to enter the highly experimental plate glass industry. He built the second plant in the country. Winslow at the age of fifty-two withdrew to attend to many other matters, large and small, public and private. So did Griswold. Chouteau and Harrison went back to the ordering of their large affairs in St. Louis. Only two men, Morrell and Holley, stayed on to carry the new process toward its greater development.

Several other aspects of the experimental period attract

attention. There is, for one thing, what may be called the general atmosphere. These people moved always in a field of resistance created by men and things. At the outset they were, as one of them said, without a lamp to their feet in a darkness where everything was without form and void. Under such conditions advances through the technical wilderness were achieved, inevitably, along false trails and circuitous routes. Consider the record of three trial blows at Troy in February 1865. "Blast 10 pounds. Blew twenty-two minutes. Vessel not hot enough. Ladle nozzle too small — 1 1/8″. 29.8% scrap. 20.2% loss. Metal came through bottom by side of tuyere, stopped it with water. Shafting of water wheel (for blower) rotten, necessitated a stop. During it vessel was lengthened 18″. The pit was enlarged. New tuyeres and nozzle substituted. Recarburizing furnace raised 6″."

Short trials and little errors; trial and tinker; trial, breakdown, change, and tinker was the way in which the machinery was slowly put together. Most of it is the story of small misadventures thrice repeated. But there were also the big occasions. Once a careless worker poured two tons of molten metal into a chilling pit at Wyandotte in which there were only a few gallons of water. The resulting explosion hurled a United States senator, who had come to observe the new miracle, across the room and blew Eber Ward out the door and on to a scrap pile. Wrote Ward to Durfee: "Come and take care of your assistant. He will kill someone by and by." The details of the technological obstacles surmounted can be safely left to the imagination, but they should be remembered. Brooding on these early years and the dreadful "fog of discouragement" that settled down over the broken parts, the unsatisfactory fragments of a finished product, and the tired men at the end of a day, Robert Hunt concluded that to bring the experiment to a successful conclusion required in those days "a faith made perfect."

And it was not only the technical obstacles that confronted the tired men. The experiments were conducted in full view of observers whose "mental capacity was suited only to ideas prevalent at the beginning of the sixteenth century." Men who managed and worked in the rolling mill at Wyandotte, men who "thought the earth was flat, the moon a fixed reflector, and that a blast of air through molten metal would 'chill her up,' placed every obstacle possible in the way of success." They wounded the spirit of Durfee with coarse comments, they conspired by trick and sabotage to slow his labors, they formed "a syndicate of sin" that actually destroyed his laboratory. To symbolize this opposition, which finally drove him from the scene of successful experiment, Durfee created Herr Unkunde Unheilswanger, who, like Satan, "exalted sat, by merit raised to that bad eminence." At Troy more elevated and informed resistance was presented to Holley. His sponsors, Griswold and Winslow, were indeed innovators, but they were also interested in calculating and limiting their risk. Where Holley sought to modify and refine both plant and machinery his sponsors sought to tie him to the Bessemer drawings, and he too was persistently besieged by doubting, curious observers in the iron trade who assured him cheerfully that the thing wouldn't work.

But the heavy fog of discouragement and resistance surrounding Troy and Wyandotte should not conceal other elements in the atmosphere. There was that "firm and unwavering faith in the undertaking they had in hand," as one man said; that "faith made perfect" noticed by another. There was also buoyancy in this air, a kind of high and wonderful excitement that is not usually associated with commercial enterprise. "First tire ever made in America by this process, Bully Boy!" is entered in the memorandum book on trial heats at Troy, while the record at Wyandotte is rendered in more stately measure, "Verily old things are passing away, and all things are becoming new."

These were no doubt simpler times; when men were prepared to be dazed by their own findings; by the perfect revolutions they would bring to pass in every iron district in the world; by the vision of themselves and their work fitted into the heroic contexts of Milton, Shakespeare, and the Old Testament. This was, after all, a great occasion, and it is interesting to see how clearly the participants recognized it.

They wrote journals and kept what they wrote. In private conversation, articles in trade journals, and in prepared essays before professional societies in later years they perpetuated the memories of those high old times. They were rather astonishingly aware of the possible influence of their work, not only in the engineering and industrial fields, but on the structure of society as well. They were lost in a furious effort to solve a complicated technical problem, but they also never quite lost sight of the superior meanings of their work, they never quite laid aside the feeling that they were caught up in a moment of great historic significance.

Robert Hunt, who became superintendent at Wyandotte for a few months in 1865, suggests all this in the following paragraph:

> While we smile over these records of a past that to some of us seems so long ago — yet in time is but yesterday — let us realize what those trials meant to those conducting them. Let us not overlook their earnest endeavours, their high hopes, many disappointments, but never failing courage. Strong faith was required, both by the capitalist and the engineer. Prob-

ably no industry ever made such gigantic strides, attained such advancement in the same number of years, as the Bessemer Process in America. But the fire that burned away its crudities also consumed great spirits. The bold investor E. B. Ward, the cultivated Z. S. Durfee; the perfect gentleman, the constant patriot, John A. Griswold . . . and saddest of all the loss of him whose hand recorded most of that which I have presented to you, the records of the actual wearing away of his great heart. Applied science triumphed, but Alexander Lyman Holley died.

In concluding this section on the experimental period one may point out that in addition to demonstrating that steel could be produced in this country by the Bessemer process, it produced two first-class ideas. The first came from Troy. At the beginning, as C. E. Dutton, a highly sophisticated geologist, once explained, "the Bessemer process was patented, and in that everything rested. Men persisted in dependence upon the converter. They said that if they could only manufacture ingots they could sell them." This was in the nature of things; the process after all was for making steel, and the center of the process was the converter. The novelty of the concept and the instrument distracted men from other considerations. They did not, however, distract Holley. As he went about building his plant from the ground up, he arrived at the conclusion that the significance of the converter lay not in its individual existence but as part of a far larger production system of blast furnaces, converters, bloomeries, and rolling mills assembled within an organized scheme. This was "the broad view of the policy of a Bessemer establishment" that Holley arrived at. The point of the system was that it closed the circle of operations that permitted the translation of iron ore into finished steel products in one place. It was this broad view that ultimately enabled the full potentials of the Bessemer process to be released.

The second idea came from Wyandotte. It was the construction of the first considerable laboratory attached to an iron- or steelworks in this country. W. F. Durfee explains the origin of the idea very clearly. "Very soon," he says, "after entering upon the study of the theory of the process (for practice at that date in this country there was none) it became evident to me that an accurate knowledge of the chemical constituents of the metals . . . was essential to its successful conduct. I reasoned . . . that while in the then state of our knowledge, it would be impossible to predict from chemical analysis just what was the best iron for the new process, it would be possible after having demonstrated by experimental working that certain irons were, and others were not, suited for our purpose, to make an analytical comparison of them, the result of which would be a permanent guide for future operations, enabling us to determine by analysis and comparison, whether any offered brand of iron was of suitable quality."

These are the words of a man who thoroughly understood not only his business, but the nature of the complex relationship between theory and practice. No doubt today they sound like a dreadful truism. But it is useful to remember that the laboratory at Wyandotte was dismantled in 1865 by a "syndicate of sin" in an act of pure vandalism; that the common practice in the early days was to send pigs 3000 miles across the ocean to be tested by trial blows in England, and that about a decade after Durfee finished his laboratory the great master of large-scale production, Captain Bill Jones, told Andrew Carnegie that chemistry would ruin the goddamn industry. Durfee's idea — against such a background — assumes new proportions, just as Holley's broad view of the policy for a Bessemer plant does when it is recognized that he left Troy in 1867 because he couldn't convince the owners of the good sense of his idea, and indeed struggled five years more before his broad view gained general acceptance.

It is possible that Durfee and Holley had their first-class unacceptable ideas simply because they were bright and because, of all those connected with the experiments, they were the only ones who had gone to college. No doubt these things helped, but there were other causes. They began in an intellectual wilderness. Durfee reports that when he set to work no man in this country had ever seen the inside of a Bessemer plant. They had no practice to observe and thus, as Durfee carefully points out, they had to start with theory. Both men, deprived of elaborate information, were also innocent of preconceptions. They were free to speculate about the place of laboratories, or to divine in their own minds the proper arrangement for Bessemer plants. It was in the experimental period that these good ideas came to the surface, in the days before considerations of common practice, available capital, competition in quantity production began to set limits on the directions a man's mind, even an innocent man, even a college graduate, could take. The problem is posed how, by education, by selection, by the creation of special conditions, such men can be left free to have good ideas after the tremendous requirements of meeting consumer demands and payrolls take over.

The implications of all this were perfectly clear to Holley and Durfee. They watched while the owners tried all the accepted methods of taking short cuts through the uncharted territory, the requests to follow the Bessemer drawings, the orders to concentrate on the converter, the importation of a French chemist, a German from the Krupp plant, an Englishman trained at Petin-Gaudet. It is quite clear that they distrusted these traditional attempts to solve the problem, representing as these attempts did efforts to stabilize a rapidly changing process at the arbitrary and obviously inefficient point of current procedure. They knew that the full development and success of the Bessemer process depended not on borrowings and indoctrination of common practice but upon a firm grasp on theory, cautious experiment, and careful analysis. Durfee tried to see that "sound principles are established in place of old empiricisms," and Holley spelled out his position even more clearly. Reflecting on his experience in introducing the Bessemer process to this country, he defined in 1876 an intellectual attitude that may sometime enter the curriculum of our schools of science and engineering. In the early years he said the steelmakers in this country worked without any real knowledge of the ores used and the products made. He still found an insufficient understanding between what he called practitioners and scientists that he found "deplorable." Many practitioners stuck to worn-out methods because of their distrust and ignorance of scientific findings. For this reason he proposed a more general education and culture for engineers and a reversal of the usual course of study. They should first acquire practical knowledge and work with their hands. "In this way they will develop a real want for the theoretical foundation of the processes they worked with and the phenomena they have observed and thus have profit from the abstract science they will study thereafter."

This statement, along with the record of his achievements, should make it clear why one of his contemporaries said of Holley, he was "the Moses who led us out of the bondage of cant and custom which made the engineer a worker only and not a thinker as well." The temptation is once again to linger with these men Holley and Durfee a little longer. But all that really need be added here is the interesting fact that neither had ever been in the iron trade: Durfee was a surveyor and architect with some engineering training at Harvard; Holley had been a journalist, an editor, a railroad expert educated at Brown. With these, by our contemporary standards, slender resources and in a short period of concentrated experiment they succeeded not only in making steel, not only in measurably influencing the ways in which steel would ultimately be manufactured in commercial quantities, but in making definite contributions to the intellectual development of American industry. These were considerable achievements.

And in truth it was a considerable time — these years that began when Z. S. Durfee set out for England in 1861 and ended when the three patent rights were put together under a trusteeship of Morrell, Winslow, and Griswold in 1867. There are to be found the satisfactions of constructive achievement, the excitement that always attends successful experiment, some very interesting and attractive men, and a moment in which the needs and interests of commerce, society, and the intellect were brought together in a single harmony. It might almost be called one of those rare occasions that reveal the possibility of developing an industrial culture as well as a society.

And like all great occasions it contains its own irony, with which it is fitting to conclude this section. It was said earlier that four men or groups of men investigated carefully the Bessemer process before it was introduced. One of them, and the first, was Abram S. Hewitt, the most progressive man in the iron trade — "restlessly alert to every new opportunity." In 1856 after reading the account of the process in the New York *Tribune,* he opened negotiations with the English inventor; he even had a trial of the process at his plant in New Jersey in the late fall of 1856. This trial or series of trials, though presided over by Bessemer's own representative, then on his way to the

United States Patent Office, were not wholly successful. Nor were the first reports that drifted over from England in the next few months encouraging. Therefore in the early months of 1857 Hewitt dropped his negotiations, and his opportunity passed. By 1866 he realized what he had missed. To cut his losses and retrieve his competitive position, he designed and built a rail with a steel top welded to an iron base. This rail never seriously competed, after 1867, with the rails that began to be produced by the Bessemer process. But in one of the interesting side effects of the innovating process, Hewitt made his own great contribution to the age of steel in this country. To get the metal for his rail he imported in 1867, the year the trusteeship for the Bessemer patents was set up, the first open-hearth furnace in this country.

Part II

PRODUCTION, PROFITS, POOLS, AND TOO MUCH PROTECTION

There remains the most interesting part of the story, the period of development. The dimensions of this period, stated in terms of chronology and statistics, are as follows. In 1867 when "the Pennsylvania people" started their new plant near Harrisburg, this plant, with Wyandotte and Troy, produced among them 3000 tons of Bessemer steel. Thirteen years later the eleven existing works in the United States turned out 1,074,262 ingot tons. In that year, 1880, when for the first time American production surpassed the British, the period of development may be considered to have been completed.

The best way, not only to enter this period, but to discover what was going on in these thirteen years is to look at the origins and development of the new plant mentioned above, the Pennsylvania Steel Company at Harrisburg. The "Pennsylvania people" who approached Winslow and Griswold to obtain a license under the Bessemer patents in 1865 had, on June 26 of that year, incorporated under the laws of Pennsylvania "the first company to be organized in the United States primarily for the manufacture of steel." They had done so because of "our desire to free ourselves from dependence upon a foreign market," and they were "determined to produce upon American soil all the principal manufactures of steel that are now at such a serious cost for freight, commission, insurance, exchange, duty etc. and often with so much delay and uncertainty imported from abroad."

Included in this group of men were J. Edgar Thomson, president of the Pennsylvania Railroad, Samuel Felton, late president of the Philadelphia, Wilmington, and Baltimore Railroad, Nathaniel Thayer of the Baldwin Locomotive Works, William Sellers, the inventor and machine-tool maker of Philadelphia. The significant figures were Thomson and Felton, and the significant thing about them in this context is that as railroad men they brought the first Bessemer plant to the iron-producing state of Pennsylvania. The maintenance costs of iron rails had persistently troubled them, the experiment at the Camden goods yard had greatly impressed them, the price of English steel rails had sorely irritated them, and the rolling out of the Bessemer rails at North Chicago in 1865 had convinced them. So they moved rapidly to take advantage of an opportunity from which the men in Pennsylvania who supplied them with iron rails drew back.

They were obviously men of considerable vision. What they had in mind was no experimental installation but an integrated production system with a total annual capacity of about 10,000 tons of steel rails. For this they were prepared to spend, originally, $200,000 on physical equipment alone at a time when Bessemer plants in England were going up for about $70,000. The original capital — 200 shares of stock at $1000 a share — was taken up "solely by parties engaged in the management of important railroads or extensive mechanical establishments." At the first meeting of these stockholders in September 1865, Samuel Felton was elected president and John Edgar Thomson and J. A. Scott, also of the Pennsylvania Railroad, were appointed directors.

Thus organized, these men approached Winslow and Griswold for a license in 1865. In January 1866, the final, formal agreement was reached between the several parties. By its principal terms, the Pennsylvania Steel Company was to receive "complete and correct working drawings of the best 'Plant' and machinery known to said parties of the first part." The company was also to have the right to send not more than two men to work for not more than two years in the Troy plant, without pay, to learn the necessary techniques for steelmaking. In addition, anyone identified in writing as a member of the Pennsylvania Steel Company could obtain free access for a limited period of time to inspect the works of the Troy plant. In return for these rights and privileges, the company paid $5000 for working drawings, a royalty of $5.00 a ton on all steel used for rails, a royalty of $10.00 a ton on all steel used for other purposes, and covenanted to keep accurate production records available at all times for inspection by the party of the first part.

By a separate and intricate series of negotiations the steel company also obtained the services of Alexander Holley to design and construct the new plant. Originally, the owners, as a "warrantee" of their conviction that "the quality of the articles they will turn out will at the start be equal and ultimately superior to those brought from abroad," had engaged William Butcher of Sheffield, England, to superintend the erection of the works. This plan was apparently abandoned when the Pennsylvania people met Holley in the course of the negotiations with Griswold and Winslow. On their part, they were impressed with his broad view of the policy of a Bessemer plant, and on his part, he leapt at the opportunity to put into practice the broad view his Troy partners were unable to accept. The terms of the agreement are not without momentary interest. In return for Holley's services, the owners agreed to build the new superintendent a house similar in size and design to one he already possessed and to liquidate the debt of about $20,000 Holley owed to Griswold, on whom he apparently had a drawing account. In return for this liberal arrangement, the company was assigned through a third person, so as not to prejudice the formal and public royalty contract with Griswold and Winslow, the right to manufacture about one blow a day royalty free.

While these negotiations were going forward, the officers of the company were not idle in Pennsylvania. After considerable investigation they chose as the place for their new works a site on the east bank of the Susquehanna, near Harrisburg "in the heart of the iron and coal regions of the State" and on the main line of the Pennsylvania Railroad. The good citizens of Harrisburg were so pleased by this decision that they took up by public collection a sum of $24,577.50 to purchase the eighty-two acres of land selected by the owners.

On this site, called first Steel Works and after 1880 Steelton, ground was broken in May 1866. In the course of the next year construction from blue limestone and brick was begun on the following buildings: a converter house 114 feet long, 100 feet wide, and 25 feet high; a rail mill 275 feet long by 92.5 feet wide; a machine shop 75 feet by 75 feet; a smith shop 50 feet by 50 feet; and two small buildings to house machines for air fans and water pumps. Projected but not begun were shops for the supplementary finishing dear to Holley's heart.

This was the basic system laid out in 1866. The whole was tied together by a network of cranes, ladles, and railroad lines along which the metal was transported from place to place in the course of its transmutation from pig iron to Bessemer steel rails.

It may be useful here to say a little more about this process of transmutation. To begin with, pig iron was brought into the converting house by small railroad cars a ton at a time and dumped in the cupola where it was melted down. The molten mass was then run off into twelve tin ladles which tipped the metal into the spouts that filled the two five-ton converters. When the charge of molten pig was in the converter, the fans were started that blew air up through the tuyeres and through the metal. At this moment the blow begins, one of the most impressive moments in American industrial history. Holley once wrote a description of the conversion process that went the rounds of the trade for a decade. It is worth quoting here:

> The cavernous room is dark; the air sulphurous; the sounds of suppressed power are melancholy and deep. Half revealed monsters with piercing eyes crouch in the corners. Special shapes ever flit about the wall, and lurid beams of light anon flash in your face as some remorseless beast opens its red hot jaws for its iron ration. Then the melter thrusts a spear between the joints of its armor, and a glistening yellow stream spurts out for a moment, and then all is dark once more. Again and again he stabs it, until 6 tons of its hot and smoking blood fill a great cauldron to the brim. Then the foreman shouts to a 30 foot giant in the corner, who straightway

stretches out his iron arm and gently lifts the cauldron away up into the air, and turns out the yellow blood in a hissing, sparkling stream which drives into the white hot jaws of another monster big as an elephant with a head like a frog and a scaly hide. The foreman shouts again, at which uprises the monster on its haunches, growling and snorting sparks and flame.

What a conflict of elements is going on in that vast laboratory! A million balls of melted iron tearing away from the liquid mass, surging from side to side and plunging down again, only to be blown out more hot and angry than before. Column upon column of air, squeezed solid like rods of glass by the power of 500 horses, piercing and shattering the iron at every point, chasing it up and down, robbing it of its treasures, only to be itself decomposed and hurled out into the night in a roaring blaze. As the combustion progresses the surging mass grows hotter, throwing its flashes of liquid slag. And the discharge from its mouth changes from sparks and streaks of red and yellow gas to thick full white dazzling flame. But such batter cannot last long. In a quarter of an hour the iron is stripped of every combustible alloy and hangs out the white flag. The converter is then turned upon its side, the blast shut off, and the carburizer run in. Then for a moment the war of the elements rages again — the mass boils and flames with higher intensity and with a rapidity of chemical reaction, sometimes throwing it violently out of the converter's mouth. Then all is quiet, and the product is steel, liquid, milky steel that pours out into the ladle from under its roof of slag, smooth, shiny, and almost transparent.

When these pyrotechnics were finished, the converter tipped its load into ladles which, carried by cranes, in turn tipped the metal into ingot molds. When the ingots were cool, they were taken to the blooming mill, where they were either hammered or rolled into blooms — finished blocks of steel that could pass more easily through the rail mill, which was like an old-fashioned laundry wringer with notches, rail size, cut in the rollers. That is the total process.

At the Pennsylvania Steel Company it took some time to put all this machinery together. The dates of development are these. The first blow in the five-ton converters was in June 1867. The rail mill was completed in 1868. The forge department, with a steam hammer to fashion the blooms, followed in 1869. A chemist and a laboratory were added in 1870. A blast furnace, to produce on location their own pig iron, was built in 1873. Three years later a blooming mill with rollers replaced the steam hammer in the forge department. In the same year the plant was employing 1500 men. It can thus be seen that it took almost ten years to bring together all the elements in the Bessemer plant that Holley had included in his "broad policy."

It took also considerable fortitude to endure the difficulties of those times, to bring the procedures within an orderly and reliable system. Bottoms fell out of furnaces, molten metal spilled all over, cranes broke down, and the quality of the steel varied widely, and often dangerously, in this period. But in spite of these technical obstacles production increased steadily throughout the period under review, as the following record of annual production indicates:

1867 — 1005 tons
1868 — 4181
1869 — 7097
1870 — 11340
1871 — 17281
1872 — 20616
1873 — 24924
1874 — 29231
1875 — 40919
1876 — 56263
1877 — 68995
1878 — 83765

The expansion of plant and production naturally required increased capitalization. Over the period of development the money invested grew from $200,000 to over $2,000,000. Of the total amount the Pennsylvania Railroad contributed over $600,000 as a corporation, while J. Edgar Thomson personally added more. The early returns on this capital were not such as to rouse enthusiasm in the hearts of the investors. No dividend was paid until 1873, and then the dividend was paid in stock. But by the end of this period, as the following figures reveal, the Pennsylvania Steel Company was satisfactorily paying its own way. Capitalized that year at $2,000,000, it paid a 20% dividend to the stockholders. Then, still having a considerable surplus, it bought in all the Pennsylvania Railroad stock at $250 and sold it back to stockholders at $100, a device resorted to because of the state tax on dividends of more than 20%. At the end of this distribution, which today must be considered generous, the company still possessed a surplus fund of over $1,000,000. In one year it had done a business which produced a net profit of almost 80% of its total capitalization.

The Pennsylvania Steel Company was the first of eleven companies built during the period of development from 1867 to 1880. One of these organizations failed, leaving, with the Troy plant, eleven companies producing Bessemer steel in 1880. They were as follows (the dates given are for the first blow):

1.	Albany and Rensselaer Iron and Steel Co.	Troy, N.Y.	1865
2.	Pennsylvania Steel Co.	Steelton, Pa.	1867
3.	Freedom Iron Works	Lewiston, Pa.	1868
		— failed	1869
4.	Cleveland Rolling Mill Co.	Cleveland, Ohio	1868
5.	Cambria Iron and Steel Works	Johnstown, Pa.	1871
6.	Union Steel Co.	Chicago, Ill.	1871
7.	North Chicago Rolling Mill	Chicago, Ill.	1872

8.	Joliet Iron and Steel Works	Joliet, Ill.	1873
9.	Bethlehem Iron Co.	Bethlehem, Pa.	1873
10.	J. Edgar Thomson Steel Works	Bessemer Station, Pa.	1875
11.	Lackawanna Iron and Steel Works	Scranton, Pa.	1875
12.	St. Louis Ore & Steel Co.	St. Louis, Mo.	1876

All these companies were of size, design, and cost comparable with that of the Pennsylvania Steel Company. Together they produced the remarkable development of the Bessemer Steel Industry in this country. It is necessary now to investigate the way in which these eleven companies sought, collectively, to deal with the problems presented to them in the fourteen years from 1867 to 1880. This investigation will deal primarily with the measures taken by the men involved to stabilize their situation, to limit or control the hazards and uncertainties with which they believed themselves confronted.

The place to start is with the railroads. No one today can quite realize the meaning of the railroads in this country from 1850 to 1890. As the first big, in fact dominating, business, they established the primary patterns, in financing, administrative structure, and pure scale, that subsequent American industrial development was to follow. In the full tide of their expansion they flooded the country with imposing new demands, social, technological, and human. By 1870 they were looked upon as the secure spine to which the smaller industrial vertebrae were tied in. It will not have escaped notice, for example, that it was the railroads that created the greatest and most obvious demands for Bessemer steel, and it was men with railroad backgrounds in most cases who took the first positive steps to satisfy these demands.

The original impetus came, in other words, not, as one might expect, from the manufacturer but from the consumer. And, as the roads had acted as the efficient cause so they continued to serve as one of the principal stabilizing elements in the development of the process. Of the eleven plants, all save one — and that, the small Cleveland Rolling Mill — concentrated on the rail business. Part of this concentration was artificially stimulated by the differential in royalties between steel going into rails and steel for other purposes. But the major influence appears to have been a desire on the part of the owners not only to satisfy an obvious demand but to achieve a degree of certainty by confining their production to a definable, known, and predictable demand.

This decision had, indeed, much to recommend it. In 1867 it was a reasonable prediction that the period of railway expansion would continue. The event proved as much. There were in that year 39,276 miles of road; six years later there were over 70,000, for an annual increase in these years of over 5000 miles. The existing Bessemer plants could not, and did not, conceivably meet the demand thus created at the beginning; hence it was natural that all energies would be concentrated on fulfilling the requirements of the principal consumer. After 1873, when the demand subsided gently for several years, supply exceeded demand, but by 1878 the rate of railroad construction advanced again and continued a gradual upward trend for the next decade. It is not too much to say that throughout this period the railways were the nurse crop for the steel industry.

A second stabilizing influence was sought in the tariff. In 1870 an impost of $28 a ton was placed on imported Bessemer rails and continued during the period under review. In 1871 the price of rails in gross tons was in this country $91.70, in England, $57.70; in 1872 the price in this country was $99.70, and in England $67.30. In these two years, even including the transport price of from two to four dollars a ton and the twenty-eight dollar duty, American purchasers could save a few dollars by buying English rails. Thereafter for the period under review this was not true. In 1875, for instance, the American rail cost about $12 less than the English rail in this country, and in 1879, about $5 less. And in the years from 1875 to 1879, virtually no rails were imported from England. Although this is in large part explained by the depression of those years, it is to the point to notice that never after 1872 did the imports of English rails represent more than a very small fraction of total American production — in 1881, for instance, about 18%, in 1884 less than 2%.

The third stabilizing influence was the Pneumatic Steel Company, or the Bessemer Association, the corporate form given to the trusteeship of Griswold, Winslow, and Morrell, which in the first instance had held the combined Bessemer, Kelly, and Mushet patents. Of this association, which combined to control the essential patents, all companies holding licenses under the patents were members. To this association every year each company contributed the royalties owed for the tons of steel they manufactured.

Now it is interesting to see how these three elements, the railroad, the tariff, and the pool, were fitted together and used as a means to manipulate and control the environment. There was first the tariff, which, high enough as it was to blot out foreign competition, established in the country a kind of antiseptic atmosphere. Within this atmosphere the eleven companies could work virtually exclusively, without diversion, at the task of satisfying the peculiar demands of one industry, the railroads. And the railroads were beautifully designed to fit into the operations of a successful pool. They figured, ordinarily, their building and maintenance programs only once a year; they could thus present the Bessemer Association with a known and predictable demand for rails at known and predictable intervals. This in turn permitted the association to apportion the annual production with pleasing accuracy among its eleven members. This possibility of fixing in advance the exact dimension of the demands together with the absence of foreign competition permitted the association, naturally, to fix, with the railroads, the price of steel rails.

Precisely how the Bessemer Association went about its business is now beyond the reach of the historian, but the main outlines of its procedure are clear enough. As the *Engineering and Mining Journal* once reported,

> Representatives of the eleven American Bessemer steel works met in private session . . . in the rooms of the American Iron and Steel Association, 265 South Fourth St. for the purpose of discussing production and prices, considering amendments to the constitution of the Association and taking action for the best individual interests. Among the representatives were: Felton of the Pa. Steel Co., Townsend of Cambria, Stone of Cleveland, Lieth of Joliet, Hunt of Troy and Wharton of Bethlehem.

It is not improbable that as these men considered the problems of production, prices, and their best individual interests, they were assisted in their deliberations by representatives from the railroads. One man, long after, remembered that

> the Pennsylvania Railroad was practically the fixer of the price of rails for a long period. The Penn. RR divided its tonnage between the 3 mills on its main line of road between the Penn. [Steel] Works, the Cambria Works at Johnstown and the J. Edgar Thomson works. For a long period rails were not purchased by other roads until the Penn RR led off with its order and after it placed its order the other roads would come in.

Again the same man reported that the eleven companies

worked together and during that early period when it was a tremendous problem to make a good rail, and, I think they fixed the price largely in conference with the railroads. I am not sure — but my recollection is that they used to talk largely with the President of the Penn RR Co. as to what would be a fair price for rails. That is my recollection. Prices changed from time to time. The railroads thought they should be lower and so they talked to the Bessemer people and down they'd come.

It was not always quite as simple as this. In the year 1877 the economy of the country was severely depressed; it was clear that the building and maintenance programs of the railroads would be considerably curtailed. Therefore, more elaborate negotiations were required to deal successfully with production, prices, and individual interests. In a series of meetings the Bessemer men drew up the following plan. First they proposed to limit total production to about 400,000 tons of rails and to assign to each of the plants a theoretical quota of about 40,000 tons each for the year. They further proposed that the price of rails at Troy should be $50, at Cleveland $51, at Chicago $52, and at St. Louis $53. Next they decided that while plants in these four cities should be held to their theoretical quota, the remaining plants would be unrestricted as far as price and production were concerned. The average price of rails in this country at the time was $43.50, or $6.50 less than the lowest price artificially set for Troy. In return for these concessions, the remaining plants would pay into a general fund $2 a ton above the usual royalty payments from which Troy, Chicago, St. Louis, and Cleveland would be reimbursed for limiting their production to 40,000 tons a year. It was an ingenious arrangement and one which suggests the considerable influence of the Pennsylvania Railroad, for the three great plants on their main line were granted the right to set production and price levels without restraint. In actual fact more heroic measures were required. The plant at St. Louis, the Vulcan Company, was closed down completely for the year in return for which it was paid out of the Bessemer fund a considerable sum, estimated by one embittered observer as about $400,000.

Other examples could be given but the point is no doubt clear enough. The railroads, the tariff, and the instrument of the pool were put together as a means to control price and production. Ordinarily this was done by persuasion or administrative decision within the group. In the hard year of 1877, Holley, Hunt of Troy, and Jones of Edgar Thomson were sent around to encourage Fry at Cambria to reduce production voluntarily. But as fears grew in the last half of the decade that existing plant capacity was sufficient to meet all the requirements of the controlled demand of the railroads, sterner measures were resorted to. At least one group of men reported to Congress that after repeated efforts they had failed to obtain from the Bessemer Association on any conditions a license to build and operate a Bessemer steel works.

Here are assembled all the dark forces of American economic history — the dominant industry controlling its supplies, the tariff, the pool. Here are also the spectacular rewards that in many a historical syllogism have been used to demonstrate that the sordid methods were consciously designed to ensure equally sordid, selfish ends of exaggerated personal gain. Consider: in the dreadful year of 1877 the Edgar Thomson Steel Company paid a dividend of 41¾%; two years later the Pennsylvania Steel Company, with a capital of $2,000,000, paid, in effect, a dividend of 77% and still retained an unexpended profit from the year's operations of $2,461,423.51. In fact, it almost seems, as one reviews the evidence, that no one in the business could lose under the system as it operated. In the year

1879, for instance, the members of the Bessemer Association paid to the railroads, their steady and virtually exclusive consumer, a freight bill of over $8,000,000.

These are indeed fair rewards. But it is not clear that the men who put together from the elements of tariff, railroads, and pool the great protective shell under which they operated from 1867 to 1881 had such dazzling returns for their enterprise exclusively, indeed even predominantly, in their minds. At the outset it must be remembered that they had on their hands a process of demonstrated value, still, however, in its experimental stages. No sooner had they begun commercial production than they discovered that, though the initial cost was perhaps not much, the cost of development was great. Changes in the design of such things as eight-ton converters, two-hundred-foot rail mills, twelve-ton ladles, fifteen-ton cupolas, and so forth cost money, and in the developmental period such changes were taking place with accelerating speed. What the Census of 1890 pointed out about the Bessemer industry in the eighties was equally true, on a smaller scale, for the seventies; advances in technique and design, upon which the survival of the industrial company so largely depended, were achieved only at great cost and at the risk that a new device when finished would, in this period of development, have already been rendered obsolescent by the creation (in the hands of a competitor, either foreign or domestic) of a better device.

In the second place, the men in the industry were confronted by a highly skilled competitor who had been first in the field, England. That country had apparent advantages of capital, trained labor, and lower production costs which could be used to undermine the whole structure of the infant industry.

Finally, these men were operating in an erratic economic

environment. The history of the country in the previous thirty years had, to be sure, demonstrated our great expansive energy, but it had also demonstrated that in a period of expansion, no one could be quite sure which new process, what new invention would pay off. It had additionally demonstrated that if in America there was a time for boom, there was also a time for bust. The conditions in this country from 1873 to 1877 could only reinforce the sad truth of these propositions.

So it seems fair to assume that what these men first sought was a degree of security for their industry and themselves, and it seems equally fair to assume that this interest in security was initially at least as much psychologically as financially determined. They were essentially administrators, presiding over the intellectual as well as the economic development of their venture. Administrators require by temperament and also by simple necessity a reliably stable environment. You can't very effectively preside over chaos. They sought by every means at their disposal to isolate and confine their universe with the thought that they could thus introduce the necessary measure of stability in their affairs. With great skill they selected the means in the environment to assist them in their task — the tariff, the pool, the steady market — and with considerable understanding they cooperated to maintain their balance in an uneasy time. Where the natural order of things was absent, they resorted perforce to artificial devices.

All this was not lost on some observers. The editor of *Iron Age* in 1879 pointed out "the advantages of artificial barriers against the promiscuous competition which results when trade matters are left in unprotected natural condition." These barriers — the tariff and the ownership of the patents — had been, said the editor, wisely used by the Bessemer men to avoid the disasters produced by the innovation in England. There, where the influence of tariff was negligible and plants could be started by anyone with money, iron rail companies first had been driven out of business with a consequent dislocation in the economy, and now Bessemer companies were turning on each other in a competitive frenzy. Prices had been driven so low that no one could make money. In this country relative order existed. The "commendable sagacity" of the Bessemer Association prevented the establishment of the two more converters that would create overproduction, the iron trade was permitted to react slowly and healthily to the gradual expansion of the steel industry, and the basis for that necessary and orderly expansion, suitably controlled, was well laid in the framework of the Bessemer Association.

It appears that the advantage of this protected little universe was that it enabled men to limit somewhat the random influence of economic forces. It also, and equally important, enabled them to control the flow of information and thus to reduce the degrees of intellectual and psychological uncertainty with which they had to contend. It must always be borne in mind that the process with which these men were involved was not only new but changing. Throughout the years from 1867 to 1881 constant modifications in procedure and design were taking place. These alterations inevitably produced a disorderly advance and poised before each owner the threat that some great change, introduced by a rival plant, might eliminate him from competition. The problem was therefore twofold: how to protect oneself from a technological lightning bolt that might annihilate, and at the same time how to distribute throughout the whole industry sufficient new information to ensure its orderly progress. The neat little world provided by patent, pool, and railroad offered a suitable solution for the problem. It gave a kind of shelter within which there could be developed what one man called "a practical fraternal feeling among the rail makers." As he explained to a Senate investigating committee, it all "began with working out of the steel rail. Various managers of works were in constant intercourse, and the owners were in constant intercourse having as the object to make with Bessemer steel, then undeveloped, the best rail possible to make. There has always been, I think, an unusual freedom of intercourse and exchange of opinion starting with that situation with regard to the rail making industry."

This was true. The cause was no doubt that no one knew enough in the early days to proceed alone. Fraternal interchange was indispensable for survival. One man, it is true, knew more than anyone else — Alexander Holley — and the career of this man is very suggestive. From 1867 to 1881 twelve Bessemer plants were built in this country. One, the Freedom Iron and Steel Works, imported English machinery in 1868 and failed in 1869. Of the eleven others, Alexander Holley designed and supervised the original layouts and construction of all but three, and in those three the influence of his thinking was apparent to any informed observer. By himself he was really the creator, technologically, of the Bessemer industry in this country. The advantages for owners in having this single peripatetic pool of information ready to hand are quite obvious. It should be equally obvious that the similarities in each plant, inevitable when each design originated in the mind of the same man, reduced to a certain extent the competitive advantage of each against the other.

From the beginning Holley went on to construct a singular career for himself. He was retained by the members of the Bessemer Association as a consultant. He spent his days in these fourteen years modifying and refining the procedures and designs and communicating his findings not to any single company but to each plant in the association. This arrangement ensured the orderly transfer of information throughout the whole industry. It also, of course, tended to limit the progress to the developing intelligence of one man, but in the existing situation this was a small price to pay for stability. And, as the years passed, steps were taken to refresh Holley's native resourcefulness. When he discovered, for his was an extraordinarily alert and realistic intelligence, that his own initiative was wearing thin, he made himself into a conduit for information from abroad. Indeed, he became a kind of confidential agent for the association. Each year in the late seventies he went abroad to examine the state of the art in Europe, dropping enough hints about American practice to receive in return full statements of European innovations. The results of his investigations he put down in secret reports for the Bessemer Association. They were amazing documents, some of them still available, filled with detailed descriptions and pictures of new practices and new devices. The nature and value of this work is well described in the *Iron and Steel Institute*. "Some of us have had the privilege of studying his annual and confidential reports on steel manufacture of the world. They were manuals of great technical skill, the results of regular visits to England and the continent prepared solely for the use of the Bessemer Association with which he had entered into confidential relations." He held "a unique position," and "manufacturers knew that if they gave him much, he brought them still more." "Every ingot," the writer concluded, "has 'Holley fecit.' "

It would be unfair to suggest that Holley made the ingots all alone. Another of his contributions to the industry was to assist in building up a school of men who could follow in his path. It will be recalled that he went from Troy to construct the Pennsylvania Steel Company works at Harrisburg. The first steel came from the converters there before the rail mill was finished, so Holley took the steel to be rolled out in the mill at Cambria. There he met the Fritz brothers, John and George, and, especially with the latter, formed a close personal and intellectual association. Hunt, then also at Cambria, regarded these two men as "among the most brilliant metallurgical engineers the world has yet seen." Between them they worked out the refinements upon which the great advances in steel production in the seventies depended. Around them gathered, in the heyday of the engineer, a group of extraordinarily competent engineers. Though Holley roved and George Fritz subsequently left Johnstown, the center for this group was the Cambria Iron Works. Cambria in the first decade became the great seminar with pupils gathered together under the discerning eye of Daniel J. Morrell. As one of these pupils pointed out, from Cambria "graduated" Hunt; Fry, who went to the North Chicago Rolling Mills; the two Fritz brothers, John who went to Bethlehem, and George; Frank Jones, who remained at Cambria; and the legendary Wild Bill Jones of J. Edgar Thomson, "the greatest mechanical genius ever to enter the Carnegie shops." These were the men who among them were responsible for the development, technologically, of the Bessemer process. In their seminar were others, less well known, less talented, but not less competent, who can equally claim a place in the great development.

They all went out from Cambria to run the plants that made up the Bessemer Association, but they retained their affection for each other, what Bill Jones called their "strong but pleasant rivalry," and, above all, their lines of communication. Recollecting the relationship that existed between five of them, John Fritz, years later, described the feeling that pervaded them all. "In the early history of the process . . . [we] would frequently . . . talk over our troubles, not high finance but the difficulties we daily met, which at times seemed almost insuperable. We did not meet as diplomats to find out what each other wanted without ever hinting of anything they wanted. But we met as a band of loving brother engineers, trained by arduous experience, young, energetic, and determined to make a success. . . . What each of us knew was common to all. Upon occasion we met at our house to talk over our troubles in detail. And they seemed so grave that some of us doubted that we could ever make the Bessemer process a financial success." Fritz concludes with the reflection that since all of the men who grew up in this period — most of them graduates of the seminar at Cambria — were "shifting constantly" from plant to plant, "this fraternal relationship was very important in the exchange of information in a new field."

The information was not always intended to help a brother solve a technical difficulty. In the early years these men sent frequent telegrams to each other announcing record heats at their own plants. The "strong but pleasant rivalry" was stimulated by cheerful messages to Cambria, Troy, and Bethlehem to the effect that the Pennsylvania Steel Company had in the past week broken previous production figures for five-ton converters. The excitement communicated itself at times even to owners. A small man on an Italian mountaintop sent up a cheer in 1876 when he heard that his plant had, for the first time, passed Cambria production for one month.

Other things than the need for information and affection held these men together. They were, they knew, in a dangerous business. In those days, said one in later years, when there was an accident in the shop — as there was once or twice a week — the question on everyone's lips was not what had happened, but who was next. These were exciting times and the men who participated in them were bound together by the recognition that they faced common hazards. It was well that they had these common bonds, for they felt frequently that they were opposed from above by owners who preferred to count on existing methods and machines rather than to gamble on the changes in practice and design that the engineers thought indispensable to progress. And from below there was the labor force, much of it in the first days imported from England and Europe, who resisted innovations because it was not so done in the old country or because each new change or improvisation brought with it the possibility of unknown new dangers in an already dangerous business.

It should be apparent from the foregoing that the explanation for the protected little universe was not exclusively economic. It set up a controlled situation in which, in times of great doubt, men could take the time to educate themselves in the nature of their difficulties. It permitted them to lay out clear communication circuits by which they were assured that all new and relevant information — on which education depended — would be channeled to them and not to hostile agents. It gave them in a time of insecurity and danger a society with which they were familiar, a feeling of friendliness. It permitted the generation of an *élan*, an *esprit de corps*, by reducing the isolation of the individual, that tempered the natural hazards and blunted the meaning of the competition they were engaged in.

From this secure and protected environment were produced not only the large profits already noted but a whole train of technological developments. In the period from 1867 to 1881 the following innovations (a partial list) were introduced in the process for making Bessemer steel: the accumulative ladle, which by accurately weighing each charge for a converter ensured more precise control of the final product; the mixer, a vat of huge dimensions which by keeping in fluid form a mixture of various pig irons could ensure not only faster production but steel of more carefully controlled quality; the detachable converter bottom, which permitted a converter to remain almost constantly in blow (while one bottom was being relined, another was bolted into the converter); the three-high blooming train; the removable shell used as a converter lining; the replacement of the reverberatory furnace by the cupola. In addition to these specific things, steady and rapid improvement was made in the organization of the plant itself: hydraulic power replaced counterweights for cranes, a method of controlling all the cranes from a single point was developed, converters were lifted high above the ground (eliminating the British system of ditches dug under the converters), rearrangements of machinery were made to produce more continuous and large-scale production than heretofore.

The sum of these improvements can best be suggested by the statistics for the year 1876, before this country had surpassed England in the production of Bessemer steel. The table is self-explanatory.

Country	*No. Converters*	*Tons Produced*
England	110	700,000
Germany	78	242,261
Sweden	38	21,789
France	28	261,874
United States	27	525,996
Belgium	12	71,758
Russia	4	8,500

As the table reveals, the United States with 11% of the world's converters produced 27% of the world's Bessemer steel. For each converter in America about 20,000 tons of steel were produced, while in England each converter produced about 6000 tons. The figures must be accepted with some reservations. For instance, it is quite probable that not all the plants worked at total capacity for the year in question, but the obvious implication of the figures is accurate. By 1876 American engineers had discovered how to make one of our own converters do the work of from three to five European converters.

Not all of this technological advance, of course, can be traced to the creation of a safe and limited environment for the Bessemer Association. And yet, in general summary of this phase, it does seem fair to reach the following conclusion. The Bessemer innovators set out quite consciously to design a system in which they could limit production to a single product, control rather precisely the amount of domestic production, order within reasonable limits the fluctuations in demand for their product, free themselves virtually completely from alien competition, and maintain a firm hold over the channels of information, including manpower, upon which advances in the means of production took place. In so doing they weathered the economic uncertainties attendant on any new venture, carrying such marginal producers as Bethlehem, Pennsylvania Steel, and Joliet through the early years from 1867 to 1874. In so doing also they preserved in safety a growing cache of information in which not only orderly but rapid intellectual — that is, technological — progress could be maintained. And finally, in so doing, by relying upon the symbolism of protection — tariff, pool, and railroad demand — they achieved for themselves a kind of necessary psychic security. The symbols, wards against external threats, permitted them to concentrate upon the internal dangers confronting them. They could develop the *esprit* and confidence to deal with present dangers, just as men may fight with higher spirit within a closed space but lose their morale before the widening limits of an unknown plain.

Considering the advantages of the system, for men in their precarious position, it is interesting to discover how hard it was for them to tell the truth about it. Indeed, they said and did all the wrong things. For instance they claimed that the existence of the tariff enabled them to put chromos on the walls and pianos in the living rooms of houses along the Allegheny and Susquehanna. They claimed that it enabled them to raise wages and standards of living. Unhappily, for the period under review this was not true. Wages for unskilled labor throughout this time remained about the same — $1.20 a day — while those for skilled labor actually fell. In 1880 skilled men in the iron trade received about $3.60; in the steel business, about $2.75. The reason for this was that so much of the steel industry proceeded by machines instead of by trained men. There is no need to rehearse here all the elaborate defenses thrown up by the steel men to protect the tariff; they were the classic syllogisms of American tariff history within which statistics, pathos, and low cunning were brought into a kind of logic. One further example, possessing virtual perfection, need suffice. In 1879 Vanderbilt ordered 12,000 rails from England for reasons, he said, "of superior quality and economy." This was not the first time he had done so; indeed the New York Central perhaps because of its geographical position had never actually participated in the development of the Bessemer process in this country. The agents of the new industry came out in full cry at the news. *Iron Age* asked what would happen to the Bessemer Association and all its dependents if everyone did this, and

Daniel J. Morrell, in a public statement, laid Vanderbilt's apostasy to three things: the "anti-American spirit that prevails in the government of the New York Central, Vanderbilt's desire to pull down the tariff structure and ruin the American Bessemer mills," and the fact that Vanderbilt was an unpleasant man. Had not old John Griswold of Troy gone to see him once, presented his card, and been greeted with the remark, "Who in hell is Griswold?" To which the steel man replied that he had called to see a gentleman, and, finding none, would retire.

By such arguments was the justification for the tariff made. The explanation for the pool was no better. Swank, the spokesman for the industry through the American Iron and Steel Association as it was then called, once endeavored to explain the Bessemer Association's contribution as follows. It had spent, he said, a million and a half dollars for the patents, and, since some of the patents had expired, the sum (which appears highly exaggerated in any case) might well all be lost. He also maintained that anyone wishing in 1880 to pay fifty cents a ton royalty and having the money to build a plant could join the association, an assertion which, on the basis of the testimony of the president of the Harrison Steel and Wire Company, is not true.

Other equally meretricious descriptions of the association were constantly turning up in the organs of the trade, but it remained for Mr. Carnegie to set it forth as a faintly religious and eleemosynary society. "I think," he said, "of some of the gentlemen in Philadelphia — Samuel L. Felton, one of the leading citizens. I think of Mr. John M. Kennedy. Those men sat there and distributed the rails among various contemporaries, representatives of the companies there, with the approval of the sellers of rails, but the moment it had occurred to them they were breaking a law, they would not have done it. . . . My early days," he continued, "were spent largely in Philadelphia, and I know our dear old Quaker friends well. There was Joseph Wharton, for instance. He sat then as a director in the Cambria Iron Co. You all know Joseph Wharton. Samuel Felton is another. Those are my fathers in Israel. That is what made me a pretty good man, just getting in association with that kind of men, and to think that they were willfully trying to violate the law — well, you could not get a man in Philadelphia to believe it."

It is just possible that the men involved, all except Andrew Carnegie, actually believed in the justifications for tariff and pool that they gave the public. One reason for thinking this might be the case is that they clearly did not understand the real nature of the environment they had created. They had built the closed system to protect themselves from outer threats; but in time they found themselves, as had so many previous constructors of self-contained societies, caught within their own protective mechanism.

They did not realize, for instance, that there was a source of danger in their exclusive dependence upon railroads. As early as 1870 the *Engineering and Mining Journal* pointed out that in Europe Bessemer steel was used in axles, boilers, and the running parts of machinery, while in this country virtually all (at that time between 85% and 90%) Bessemer steel went into rails. In 1875 Alexander Holley returned from one of his roving commissions to give a speech before the American Institute of Mining Engineers. He was aware, he said, that American Bessemer manufacturers were thinking about reducing production for fear that present rail demand would not come up to the capacity of their works. A wiser course, he urged, was for the manufacturers to experiment with making other grades of steel for other purposes instead of devoting their time to schemes for limiting production. Europe, he pointed out, used Bessemer steel for all kinds of products. In America, the market for steel buildings, steel bridges, and steel vessels was opening up, but our men "have not yet made any study of the use of steel and are therefore afraid of it."

These warnings and others like them that appeared in increasing volume after 1876 went unheeded. The dependence on rails continued until, near the end of the decade, even *Iron Age* was calling the blind allegiance of the association to the railroads "a marvel of stupidity."

Even in the rail business the regrettable effects of a closed circuit of communication were made manifest. Not even the energy and intelligence of Holley could feed the whole industry indefinitely. The unhappy fact of the matter is that although this country by constant improvement in the technique of production rapidly met and then surpassed the current demands for rails, it never, in this period, made a very good rail. A decade earlier the English had hurt the sensibilities of our manufacturers by calling the poorest grade metal in their yards "American rails." In the seventies the situation was not as bad, but the fact remains that our rails were always then inferior to those of Europe. Vanderbilt knew this when, for reasons of economy, he bought his material in England. The cause for this was that concentrated as the plants were on quantity production, little attention was given to improving quality. As Bill Jones said to Mr. Carnegie in 1876, chemistry might ruin the goddamn industry, and in the fierce competitive struggle for production that existed at the time, he might well have been right. In any case, the urge to investigate was shut off, and the American rail remained inferior.

An interesting discussion about this took place in 1876. The Iron and Steel Association took the view that the way to improve the American rail competitively vis-à-vis England was to raise the tariff. In rebuttal, the *Engineering and Mining Journal* said the only way to secure competitive advantage was by the development of fundamental science. Thus, in a way, the tariff became a shield for ignorance. In the period under review, no single fundamental contribution to the manufacture of steel was made by any American. Incredible feats of refinement of existing procedures and machines to produce were accomplished, but all the great metallurgical advances including the lifesaving Thomas Gilchrist basic process were introduced from abroad.

In yet another way the Bessemer men were trapped by their own obscurantism. *Iron Age* in 1879, commenting on the great Bessemer year that had just passed, pointed out in passing that the Bessemer Association by its existence did tend to drive other men and money into the open-hearth process. At the time the open hearth was little more than a small cloud on the horizon, no bigger than a man's hand. In the year before, 1878, the first rails had been made from open-hearth steel — 9397 tons in all. The same year over 700,000 tons of Bessemer rails had been manufactured. Fifteen years later the open hearth surpassed for the first time the converter, and since then has never been superseded. The Bessemer Association did not achieve all this singlehanded, and in time, no doubt, the same result would have been achieved because of certain advantages in the open-hearth process, although even today in Europe Bessemer steel annually exceeds steel from the open hearth. But the association, by refusing to enlarge its plant, did certainly accelerate the development of the process that ultimately reduced the converter to a minor consideration in the industry.

There was another, and final, defect in the structure devised by the Bessemer men to control their environment. This defect was called to their attention by Andrew Carnegie. As everybody knows, Carnegie, in the years immediately following the Civil War, had been nimble in the selling of railroad bonds to the iron trade. At the beginning of the seventies, as head of the Keystone Bridge Company, he was something of a figure in Pittsburgh. With his interest in iron and his close association with railroads, it was natural to suppose that he would have been early attracted to the new process for making steel rails. Such was not the case. While others in the state were moving out to take advantage of the developing opportunities, Carnegie held back. His early reluctance was supposedly dissipated by a Pauline conversion induced by his first view, about 1870, of a Bessemer plant during a blow. This thing, "half a furnace and half a cyclone," is reported to have charged "the sudden blast of his ambition and resolve." It blew away the sordid lures of high finance and stamped steel "upon his mind with a white-hot impress."

The little man's decision to enter the Bessemer business seems, in fact, to have been the result of a much more deliberate calculation. As he warned his friends, "Pioneering doesn't pay a new concern. Wait till the process develops." Years later he recalled how he had watched his old friend John A. Wright, "one of the best and most experienced" men in the iron trade, president of the Freedom Iron Works, lose his substance as a pioneer. Wright went to England, bought a converter with supporting machinery in 1867, and set it up at the ironworks at Lewiston, Pennsylvania. The capital required had been underestimated, experiment proved more difficult, the firm failed. Watching his friend Wright, the early tribulations of the Pennsylvania Steel Company, and the troubles at Troy,

Carnegie decided to wait. By 1872 Carnegie concluded that the time was ready to move into the new industry. With his partners, Phipps, McCandless, Kloman, and the rest, he gathered together $700,000 and set to work to build at Bessemer Station, Pennsylvania, the great Edgar Thomson plant. There was a little preliminary trouble about the name. The president of the Pennsylvania road told his old friend he didn't want his name associated with a plant making steel rails because the quality of United States rails was so poor. A little persuading did it, and so on August 25, 1875, the Edgar Thomson plant made its first blow.

Carnegie had waited until the process proved itself; he had selected Alexander Holley to design and build the plant, after Holley had built five other works; he had hired from Cambria the well-trained and extraordinary Bill Jones. The care with which he proceeded justified itself. The profits of the new plant in its first four months from September 1, 1875, to January 1, 1876, were $41,970,06.

Thus equipped with the most modern plant, a tested procedure, thoroughly trained men, and a going concern, Andrew Carnegie turned to disrupt the little world of the Bessemer Association. Such things were done to the fathers in Israel as had not been done since the days of Old Testament vengeance. By chance, he once told a friend, he went to a meeting of the Bessemer Association after he had been invited as president of the Edgar Thomson. They met in Philadelphia, and he sat amid "the grave chin whiskered Quakers" mapping the year's campaign. He watched as they allotted the year's rails to various companies — Cambria, 19%, Pennsylvania, 75%, and down to Edgar Thomson, 9%. At this figure Carnegie sprang up to claim a share as "large as the largest." The grave Quakers laughed, but Carnegie brushed them aside with the claim that he could roll steel rails at considerably less than any of them. "If I don't get as high as the highest," he said, "I'll withdraw and undersell you." To his proposition all then agreed.

The situation created by the steelmakers for their own security was beautifully designed for the success of Mr. Carnegie. For ten years before he entered the industry, the pioneers had painfully constructed a restrictive system which would enable them in concert to limit and control the uncertainties, financial and technological, that confronted them. In their zeal for security, they had forgotten or overlooked the fact that the very instruments they had designed to give them control over the whole situation could be converted easily into dangerous weapons against them if someone of brains and daring himself seized control of the instruments. Andrew Carnegie was a man of brains and daring.

Much has been said and written about this perfectly extraordinary person. A good deal of what has been said and written seems only to conceal his quality, and none of it really provides a sensible guide through the labyrinth of his personality. It is clear and agreed upon, however, that he was a wonderful salesman and a wonderful judge of men. Much of the success of Edgar Thomson depended on these two assets. Not so obviously, but as certainly, he had what he took great pains to conceal — the ruthlessness of a great general. He could with disarming glee dance upon a mountaintop when he heard his plant had exceeded the production of Cambria, but he could also write his general superintendent, "If we are to meet [a year of low prices] rightly, the track will be clear after the war is over — one year without a dividend on Cambria, Penna. Steel, Joliet, and North Chicago will make some amicable arrangement possible — bear in mind that with Cambria secured we could today have fine prices in the west." "Joliet," he continued, "is in a death struggle [the time was 1875, two months after Edgar Thomson went into production] and owing to Meeker's commissions will take orders at cost. Having faith in our ability to manufacture cheaper than others I do not fear the results of a sharp fight."

He could also write a year later the same superintendent, Shinn, that he wanted *"all"* of him for Edgar Thomson. "I want somehow or other to get you root and branch — compensation can be arranged — I don't care about money so much as . . . success."

The tone is not quite that of Bill Jones's "strong but pleasant" rivalry among his brother engineers, but the intent was not dissimilar. What Edgar Thomson wanted, from senior partner to shop foreman, was . . . success. No wonder the brooms were hoisted on the converter house week after week as the production records were swept clean away and the cost of production at Edgar Thomson went steadily down.

Other things Carnegie also had. One was an eye for significant technical information. This man, described often by his own associates, and men who later listened to them, as the one who did the advertising and drove the band wagon, could write that in a year of low prices, rerolling was most important: "Jones can't do it successfully without new rolls and a shear — with them he can beat Cambria badly." And again, "I see it is done by a lime lining in the converter which is found to absorb the phosphorous. Dr. Siemens is out in a letter saying he had tried that for his open hearth furnaces, and found it could not be made to stand; this thing appears to have been overcome . . . it is too important not to invite our earnest attention — what a failure for Cambria! We should, I think, profit next largely, while Chicago and St. Louis would not be much advantaged. . . .

"I shall of course keep you informed. If we could go to 13 in phosphorous and take even a little out in conversion we should have Republic."

But the real success rested primarily upon his sense of situation, his understanding of what was required and his ability to supply it. It is apparent in what would now be called the pure corn of his testimony before the Stanley Committee with his talk of dear old Quakers and his offer to give the chairman, lacking both in information and money, a library for himself. It is apparent in his selection of precisely the right moment to enter the Bessemer industry, and it is again apparent in his perception of how to proceed once he was in the business. As he told his partners in the sixties that pioneering didn't pay (agreeing, no doubt, with Reginald that the first Christian gets only the largest lion), so he told them in 1874 that the time was ripe "to put all your eggs in one basket — and then watch the basket." This was his own great innovation, and it is a perception of considerable originality. The drift in the industry was the other way. By 1875 men had come to recognize clearly the essential ingredients of the steel business: coal and coke, ore, transport, the converter, and the fabricating machinery. An urge to put all these things together developed in the owners after the success of the converter had been demonstrated. At St. Louis, the Vulcan Company was formed to put together an ore field, a coal mine, a railroad, and a converting plant. This was the principle of consolidation by vertical trust. It conformed to the pattern of most of the men's lives — men like Ward, Harrison, Felton, had been in all kinds of industrial enterprises. Carnegie in 1875 would have none of this; he put his eggs into the single basket of steel, watched it, increased the heat with his own communicable intensity of spirit, and hatched the modern industry. Later, around the secure base of steel, he acquired ore fields, sources of supply for coal and coke, and shipping lines, but not until his base was secure. And in the meantime, the elaborate structure of Vulcan disintegrated because with its manifold obligations and interests it could not manufacture steel as cheaply and as well as could Andrew Carnegie.

There is really nothing more to tell. When Carnegie had established the upper hand, as he had by 1880, the development stage of the industry was over, and the essential patterns had been set. He went on from strength to strength, acquiring Homestead, Duquesne, the H. C. Frick Company, and with it the general staff intelligence of Henry Clay Frick himself. The final projection of Carnegie's pattern was inevitable — the United States Steel Company, where all the eggs were in one basket, or at least a lot of eggs. These are simply refinements of the major premise. Long years after this interesting early period, Carnegie confided to an investigating committee that he hated pools and had never attended one; in his mind the meeting in Philadelphia of the biblical characters was really not a pool. In a way all this is true; he had perceived that he arrived at a time when one man by skill and perseverance could take over a whole system. He entered the Bessemer Association once to gain control over the carefully devised instruments that enabled him to rule the whole domain. These instruments, used by his predecessors who had created them to limit the dimensions of their enterprise, he used to organize and expand and dominate the entire industry. He became himself the pool.

There remain only some summary reflections. The first has to do with the counterpoint, appearing at the very outset and continuing throughout, between knowledge and practice. As one man said in 1890, "almost no chemical knowledge was required. A good mechanical engineer to build, a good administrator to manage — to select his men with judgement, to treat them with gentle just firmness who can appeal now to pride, now to cupidity, identify the interests of the man and masters . . . whose deft velvet-gloved iron fingers can hold a thousand reins." This is an accurate description of a state of mind; but of course it is not true, as the men who held the reins were constantly discovering. They would hire their good engineers to build and their good administrators to manage, and then find that they would have to get more information in order to proceed. So men were imported from Germany, France, and England to introduce new ideas. When the usefulness of these ideas was exhausted, the search for more ideas and new men went on. For a time the industry lived pretty much off the wits of Alexander Holley, who himself had to go regularly to Europe to refresh his own creative intelligence.

It is true enough that the development of the Bessemer process is a triumph of American know-how, but it is equally true that this know-how rested upon information imported from elsewhere, and that the know-how was refined only as the basic information improved. In other words, our procedures rested on no very secure native intellectual interest. They were advanced on a kind of *ad hoc* basis. When existing practice proved inadequate, we went out and borrowed a particular idea that would remove the inadequacy. There was no general advance in understanding. Thus while we greatly increased the production for each converter, while we greatly refined the organization of plant, while we introduced countless useful modifications in design and succeeded in producing many more rails than could be used or bought, we failed to advance the general understanding of the process. From Europe came the great saving operations for the industry, the information obtained through spectroscopy of what took place in a converter, variations in the quality of steel produced by varying elements in the pig iron, the basic process of Thomas Gilchrist, and the new departure of the open hearth.

The counterpoint between knowledge and practice as revealed in this situation represents a great success in the engineer's objective in the economic use of energy. There is a kind of elegance in the way, without wasteful excursions into the realm of theory and speculation, we reached out to select exactly the kind of information we needed to solve particular problems arising at the several stages of development. When no one knew much, or enough to survive, Durfee built his little laboratory; when the problem of survival involved increased production and not additional information, chemistry, quite correctly, could be conceived of as ruining the goddamn industry. By introducing new factors it could dislocate the orderly means of production. So rails, even though they were rather poor rails, were turned out in staggering quantity. But successful though the method was, it need not conceal that the indispensable and marvelous know-how rested, as it must always rest, on a continuing supply of ideas. The production of ideas, contrary to the mere selection of the appropriate ideas from an available pool, cannot be achieved with great economy of energy and elegance, because it requires investigation and thought, and thinking is a wasteful process. It was natural then to concentrate on method in a fiercely competitive time and to leave the ideas to others.

A second reflection has to do with the excitement, or

perhaps merely the vitality, of these times. John Fritz believed that there was no one who entered the industry in the early years who did not at some point heartily regret it, and this is probably true. But offsetting the momentary irritations and fits of despair there was the great undertow of fascination and excitement that comes through the words and memories. The spectacular nature of the converter itself, the drama of lightly lifting and controlling burdens of ten-ton weight, the brilliance and beauty of the blow, the constant hazards and real physical dangers of the calling all conspired to attract these men. It is amazing to find how often they set down, in autobiographies, memoirs, professional journals, and memorial volumes, their vivid recollections of these times. They were a part of something historical in its contributions, and of something destructive in its newness. They were proceeding into the technical wilderness. Naturally this expedition appealed most directly to the engineers. They were confronted by fascinating problems of development which with both skill and daring they proceeded to solve. It is interesting to notice here in this connection how high they rose, in salary and status, in this period. Bill Jones was, to the amazement of all, paid more than the President of the United States, though something less than Andrew Carnegie, and in all plants the good engineers were very truly sought after. He also had a clear voice in policy, for upon what he could do or convince others he could do, the determination of policy rested. After the years of development, the great names in operations — Jones, Hunt, Fry, Fritz, and the rest — drift slowly out of the picture and, no doubt, down the status scale.

A third reflection has to do with the persistence of national characteristics. Where the British, for instance, tended to continue investigation of chemical structure and exploration of manifold uses for the new product, the Americans concentrated, as soon as it was discovered that the process would work, on means to increase production. This American approach is revealed in the steady development of bigger machines and laborsaving devices, but also in the prevailing attitude, in what the nineteenth century called "hustle." A British traveler, awed by the size and complexity of the Pennsylvania Steel Company, said he would like nothing better than to sit down on an ingot and watch all day, to which Holley retorted that he would have to go back to England to find an ingot that had been left alone long enough to cool off enough for sitting purposes. It may be worth noticing here that this differential in national characteristics is not necessarily, or even probably, the result of some difference in the national gene. More certainly it derives from the conditions presented, and in this country the central condition of the nineteenth century was the need for large-scale production to fulfill the requirements of a large, underdeveloped country with a growing population.

These are, perhaps, reflections of secondary significance. Two others may have more importance. The first has to do with men. As already noticed, the introduction and development of the Bessemer process attracted a considerable number of men, able and distinguished in a number of fields. Of these men, three seem outstanding, indeed remarkable — Eber Brock Ward, Alexander Holley, and Andrew Carnegie. Appearing at different stages in the development of the process and rendering different services, they nevertheless among them can claim the primary credit for the deliverance of the steel business to the American public. Ward got it started, Holley brought the technology to a high and practical level, Carnegie provided the ultimate administrative solution. Now the thing that is interesting about them collectively is that although they were widely different in personality and mentality, they shared one thing. Each in a sense lost interest in his own creation, or perhaps it is more to the point to say that the original interest of each was displaced by another. Ward, as has been noticed, left the Kelly Pneumatic Process Company to continue with his other large concerns and to explore the new possibilities of plate glass. Holley, by the end of the seventies, was turning more and more of his attention to the development of the open hearth. Shortly before he died in 1882 he told a friend that he regretted he could not live to leave to the American public the perfected Siemens-Martin process. Similarly with Carnegie. As early as 1889 he began to think about selling off his vast imperium, and after the consolidating and organizing intelligence of Frick took over in 1890, he paid less and less attention to the works, until they were finally sold in 1901. He had told Shinn in 1876 that it was not money that interested him — which was a shame since he had so much of it — it was success that was his true objective. By 1889 he knew he was successful, and only once again in his career was he fully roused to the interesting possibilities of the steel industry. That was in 1900 when he derived infinite delight in devising a series of masterful measures to drive the price of his holdings up before selling them to J. P. Morgan.

It may be said that of all those involved in the introduction of the Bessemer process, these three were the complete innovators — the men with the capacity to perceive from a variety of choices the particular meaning of a new idea or device, with the desire to commit themselves sufficiently to the idea or device while it is in the developmental stage, and the sense of timing and individual freedom to move on to other things when the idea or device has been brought to perfection or to a point where it may be generally accepted. The key to the character of these may well be that they were not attracted to the new element — whether idea or device — for the thing itself, nor did they become irrevocably committed to it or its development; they did not, in other words, become personally involved with its continued existence. Compare these men with another able man — D. J. Morrell of Cambria. He was early attracted — from the moment of Kelly's fireworks in 1857 — and he remained fixed in his attachment through times both thick and thin until he died. He gave evidence at all times of excellent perception, fine judgement, and real fortitude. But, remaining as he did fully committed to the Bessemer process, he got, in the later stages, bogged down with defenses of pool and tariff and the maintenance of existing conditions.

The difference may be that where Ward, Holley, and Carnegie were creative personalities, Morrell was not. To these three the particular medium, Bessemer steel, and money were of secondary interest, but the medium gave them the opportunity to express themselves in new ways, to make new and exciting arrangements, to create something out of their thoughts and feelings that had not existed before. Once achieved, they were readier to move on into other mediums that gave them renewed opportunities for creation — plate glass, open hearth, peace. They did not identify themselves with their creations or find satisfaction so much in continuing the artifacts they created as in the act of creation itself.

And finally, there is the reflection that the introduction of the Bessemer converter from 1857 to 1880 was an exercise, not perhaps in timidity, but in extreme caution. For ten years after Kelly had first developed the idea, no one paid any attention to him. It took seven years after

Bessemer had read his paper, and at least six countries in Europe had acted on it to build converter plants, before an experimental works was set up in this country. It required a consuming need in the railroads to kick the manufacturers into production. Once in business, the producers did everything in their power to restrict the use of their product to one purpose, and tried hard, when that purpose produced a contracting demand, to limit by arbitrary methods the growth of their industry.

These are the main lines of the early development. Subsidiary data only delineate more clearly the principal contours. The principal manufacturers actually risked little financially in the development of the process; only one man, Wright of the Freedom Works, was really ruined. The others invested, ordinarily, only small fractions of their holdings or waited until their ironworks profits enabled them to take a small flier into building auxiliary steel plants. They protected themselves further by making use of the corporate structure with its limited liability. They discouraged the impulses of the technical men to introduce, beyond the point where they could obtain an immediate and temporary advantage, technical advances. They restricted current information and prevented the development of sources for new information with all the obscurantism of the absolute monarch uncertain of his throne. The list could be multiplied almost indefinitely.

It should now be said that this extreme caution was for a major part of the years from 1847 to 1880 no doubt justified. One should, in casting up accounts, consider the very real difficulties and dangers facing the men who introduced the process; one should also consider the number of apparently good ideas that in practice failed to prove out in the period — indeed, in any period. It should be recognized, and may be agreed upon, that the men involved, in view of all things, including human nature, were in their caution wise and sensible in their own generation.

The trouble is they outlived their generation; that is, the conditions in which they had set up their elaborate protections against uncertainty suddenly changed about 1874. At that time the methods for quantity production had been perfected; the need for the product in the whole society greatly exceeded the demand provided by the exclusive customer; the organizational structure of the industry had been pretty well stabilized. The time was ripe, in other words, to cast off the artificial restraints, to exploit the new demands, and to enjoy the innumerable benefits of free enterprise. Shut up in their own enclosure, living off the information of a decade earlier, wedded to their own small concerns, and bemused by their investments of time, money, and pride in the converter, they could not discern that the changing times required new solutions; they could not even discern that times had changed. They had brought, by 1874, their own machinery of protection and restriction on to a dead center.

From this it was rescued, in a sense by an outsider, by one willing to wait while others rushed in for patent rights and first claims until the industry was ready for a new move forward. So Carnegie, with his sense of timing and situation, came forward to usurp the throne in the state other men had created. His action was not unlike that of Theodore Roosevelt in breaking the deadlock that existed in the Navy in 1902 over the new system of gunnery.

In sum, there are more stages in the innovating process than are at first apparent. There is the inventive stage, the stage in which the invention is applied by the first entrepreneurs, the stage in which other entrepreneurs and engineers refine and consolidate, and the stage in which still other entrepreneurs take over to expand. At each stage new manpower appears to displace those who lost their interest or their hold. Apparently only those outside at each stage have the capacity to gauge the state of things within, and the will to accommodate themselves to that state.

8

Some Proposals

A little more than a century ago, Matthew Arnold asked what coal, iron, and railroads had to do with sweetness and light. The answer often given over the last hundred years appears to be, on the whole, "not much." The dark satanic mills don't seem to grind out the stuff of the New Jerusalem. Many people, confronting the question posed by Arnold, have reached this same conclusion in quite different ways. For instance, some like Albert Jay Nock have all along wanted to close the mills down and revert to times more pastoral; some like Robert Dale Owen or William Morris have offered diversionary, small-scale modifications. And some, accepting the condition as a given, have produced the foreboding myth or cautionary tale, all the way from *Frankenstein* through *R.U.R.* to *Brave New World* and *1984*. The message of all these witnesses seems to be, one way or another, that what is supposed to be humane about humanity gets ground up into smaller pieces as the horsepower in the engines increases.

Some others have admitted the condition but accepted it with greater equanimity. They are the ones who come down hard on the side of the general, steady rise in the standard of living. Machines wherever used over the last hundred and fifty years may not have let in much sweetness and light, but they have steadily made the physical facts of life much more tolerable. This position, which has much solid evidence to support it, is usually held by those who tend to believe that human experience is not so much a matter of unheard melodies as of blood, bread, and money.

The sum of all these varying attitudes has been a persistent allegation that what men did with their engines and powerhouses had very little to do with what Matthew Arnold, in another phrase of his, called the best that had been thought and said — and felt — by human beings. There has always been a small school of thought with a somewhat different view of things, a number of men who have tried to establish some sort of connection between sweetness, light, coal, iron, and railroads. In quite different ways Henry George, Graham Wallas, and even Max Weber may be said to have taken on the job. The most thoroughgoing and by far the most influential of this school was Karl Marx. Holes have been found in his economic theory, and some eccentricity appears in his definition of sweetness and light. Furthermore, his scheme in practice has tended to produce weird results. To maintain his argument in the actual course of human events it has been found necessary, as so often before from the *auto-da-fé* to the fires of Smithfield, to turn many of the human hopes and desires his scheme was supposed to take into account into ashes.

But neither the intellectual limitations of the Marxian system, nor the fact that the system was conceived less in affection than in indignation, nor yet the fact that it doesn't seem to work very well should diminish the significance of the effort or intent. Marx was out after a plan that would fit the persistent human energies and desires into the developing mechanical and economic energies of his time. His failure may have been that in discerning some of the most important elements in the human condition, he thought he had discerned them all, but in failing, he established some very illuminating connections between man and machinery. One does not have to be a Marxist to have sympathy and respect for the aim any more than one has to be a Victorian to get through the regrettable cambric phrase to the point Matthew Arnold was making.

They were all — George, Marx, Arnold, and the rest — in one way and another making somewhat the same point. None made it more bluntly than Thomas Huxley. He came to Baltimore toward the end of the last century to say that he remained unimpressed by all the power, natural resources, knowledge, and machinery that had so greatly extended man's competence over his physical environment. "The great issue," he went on, "about which hangs a true sublimity and the terror of overhanging fate is, what are you going to do with all these things?"

The reason for bringing all this up is that after more than a century of speculation, premonitory myth, logical system, sentimental repining, legislative action, and social accommodation, the initial question appears to remain. If anything the apocalyptic vision of Mrs. Shelley and Huxley's terror of an overhanging fate seem by events to have been supplied with much raw substance. What we can now do with all these things, among other possibilities, is to blow the whole world up. But we can also, as T. S. Eliot said, less dramatically go down the drain with a whimper. The system of ideas, energy, and machinery we have created to serve some essential human needs, it now appears, may, if not sufficiently tended, shrink human beings to the restricted set of needs the system was designed to satisfy. Or, to put it another way, the system may have acquired a mass and scale and intricacy and internal rate of change that make it increasingly difficult for human beings to live comfortably and fully within it. Or, to put it yet another way, we may be caught in the irony that at the very moment when by our wit we have developed the means to give us considerable control over our resistant natural environment we find we have produced in the means themselves an artificial environment of such complication that we cannot control it.

It is obviously difficult to state the case without a grandiloquence which then begets the grandiloquent reply—the dreadful image, the revolutionary program. These are, in fact, large matters, and large consequences are no doubt attendant upon them. Yet here is not the place to make a trial at great summations or at all-inclusive propositions. What follows are modest, tentative proposals. Take them not as a systematic analysis of a great theme, but for what they are, occasional reflections, some already quite familiar, on certain modes within the theme.

To state the case in its simplest form, the current problem is how to organize and manage the system of ideas, energies, and machinery so it will conform to all the human dimensions. Experience suggests first that there are three things not to do. In the older, simpler days the solution, often, was simply to burn down or break up the new mechanisms when they introduced distortions in settled customs. This is what C. P. Snow has called the Luddite approach. This primitive response produced, historically, only temporary and local accommodations, as in the ancient case of the Pythagoreans who attempted to preserve the purity of their mathematical expressions by putting to death the man who discovered incommensurables. Whether it appears in the form of an angry workman dismembering a loom or as a fireman in a cab without a firebed — which is a kind of symbolic destruction — it does not get to the heart of the problem. Indeed, it merely complicates it.

Nor will it do to try to arrest the movement in the system that produces new ideas and machines by adopting a modern equivalent of the direct and literal-minded attack of the Luddites — that is, by burning down or cutting up the budgets of universities or commercial laboratories from which the novel thoughts and applications so often come. In a world in which the trained intelligence plays a continuously larger part, conditions to sustain the lively mind must be preserved. If a man finds evidence that the neutrino may exist, he will have to search for the neutrino; if he thinks of the possibility of the transistor or the fuel cell, he will certainly have to try to find them. Besides, the inertia in the present scientific technological system is too great to end by artificial means. One thing, one series of findings, sets in train the events that produce another thing, another series of findings.

Nor will it serve the purpose to develop some grand synthesis to bring the whole system of ideas, energies, and machines under suitable control. For one thing, such grand syntheses — whether Platonic or Marxian — never work out very well in practice. There are, apparently, too many variables in any situation where human beings enter in to be successfully comprehended in any fixed grand design. And in this particular situation, which is dominated by accelerating change, any grand design would soon become quite obsolete.

So one must look elsewhere to find the means to govern the system in our own interests or in conformity with all the human dimensions. Most simply stated, the problem is, first, how to reverse the process, noted in the essay on the *Wampanoag,* by which the technology tends to create its own environment and set of conditions. Put even more simply, as the mechanism steadily increases in power and scale, the tendency is to fit men into the machinery rather than to fit the machinery into the contours of a human situation.

One way to begin thinking about this problem is to creep up on it out of the simpler past. The introduction of the pasteurizer in the last century may supply a suggestive point of departure. It is an intricate historical anecdote containing many tangled threads that in the unraveling would require much time and more research. But some of the main lines are clear enough for the purpose here. Pasteur developed his idea in the 1850s while working on the diseases that had attacked the vines in the French wine country. The first general application of the idea appears to have taken place about a decade later in Denmark. The Danes made a considerable part of their living by exporting cheese. They discovered that they could measurably increase the uniformity of their product by pasteurizing milk and reinvesting it with small colonies of bacteria. Accordingly, the developing cooperatives began to require the process for all cheeses made for export. It was then further discovered that calves drinking the pasteurized milk developed bovine tuberculosis far less often than calves raised on raw milk. Accordingly, laws were passed requiring the pasteurization of milk fed to the stock in commercial herds. That people might similarly profit from the use of the process was not then thought of.

In Germany somewhat later, it all began with beer, when the brewers found that they too could ensure a greater standardization of product by using the process. In time —

toward the end of the century — German chemists and medical men began to accumulate evidence that raw milk often carried bacteria harmful to people as well as to beer, wine, and cheese. Pasteurization of the milk sold by commercial dairies then became a requirement imposed by the German Emperor. When the idea crossed the channel into England in the last years of the century, interest was sufficient to put the matter before the voters in various communities, but for the most part, the voters proved apathetic when they were not positively resistant. The thing in these years never caught on.

In America the case was different. Pasteurization was introduced in quite a few city dairies in the late eighties and early nineties clandestinely. In these days the milk sheds were moving farther and farther away from the urban centers, and in the course of the extending transport from farm into the bottling plants and out to the milk routes, the milk tended increasingly to sour. So as a means to preserve the life of the milk the dairies began to pasteurize. They did so secretly because the citizens had a natural resistance to drinking milk with, as they often said, dead bodies in it.

This resistance was overcome in the first instance in an interesting way. A wealthy New York German Jew on a trip back to the old country in the nineties heard about the new process and its possible useful effects. He determined to introduce it upon his return into a depressed area with which he was thoroughly familiar because he had once lived there — the Lower East Side of New York. So one fall he opened a free milk kitchen dispensing pasteurized milk for all who came. Few came. But he was not only a generous man and persistent; he was also clever. With the aid of a biologist at Columbia he made two blocks of the Lower East Side the site of a considerable experiment. In one block the residents were persuaded to get their milk from his free kitchen, while in the other they bought as before unpasteurized loose milk from the itinerant dealers. Over a period of several months a careful count of the appearance of scarlet fever and diphtheria in the two blocks was maintained. The final reckoning produced a rather dazzling demonstration of the effectiveness of pasteurizing in reducing the incidence of disease. It was not long after that the process was required of commercial dairies by New York City ordinance, and quite rapidly after that other cities in the country passed similar ordinances.

In the following decades of the new century the idea of pasteurizing milk went forward unevenly but progressively until it became of fairly standard practice, except of course with the French.

These were, as has been said, older and simpler times, and the pasteurizer is not a very powerful piece of machinery. But the history may be taken at least as a kind of parable. It was brought to mind two years ago when a freighter entered the harbor at Beirut. There thirty years before had been a small port reeking of the Middle East. Now the good people of the city, having discovered the miracle of poured concrete, plate glass, and the internal-combustion engine, were busily at work constructing the shape and meaning of Miami Beach among their lovely hills and water.

There is no cosmic proposition intended by these juxtapositions. Beirut and Miami Beach are really only a diversionary bypass on the way to the point of the parable of pasteurization. The pasteurizer was not a powerful piece of machinery, but its history suggests the possibility that mechanical applications can be modified, one way and another, by the pressures of the customs of the community.

What it became and how it was used were functions of the cultural needs and attitudes of the various different societies that used it.

Contemporary machinery is far more elaborate and powerful and pervasive in the social fabric than it was in the days of the pasteurizer. The cultural constraints of today seem, on the whole, far less binding on any society than half a century ago. Yet if one went about it in the right way, it seems still quite possible that a new kind of culture could be built up that would contain the new technology within appropriate limits.

And the creation of such a new culture would seem to be a first order of business. At a time when we are developing the capacity to do almost anything that occurs to us, from going to the moon to transplanting one man's brain to another man's head, new means to select and judge among all the things that are possible must also be developed. If we are to manage the powerful system we have created in our own interests, we must also create a new sort of culture that will give clear definition to what, in the new scheme of things, our interests really are.

One of the difficulties in the past hundred years has been that technological advance took place within a cultural surround that had been designed over centuries to deal with situations in which there had been only minor or low-powered technical systems, and such as there were remained reasonably stable for long periods of time. Modern technological advance therefore increasingly took the form of systematic destruction of customary attitudes and values that had been created, often unconsciously, to control quite different conditions, conditions in which men were more troubled by the power they did not have than by the excess of power at their disposal. It is quite possible therefore that the celebrated war between the two cultures has taken place not over some fundamental cleavage in human understanding but over something else. On the one side there have been men anxious to change conditions by expanding man's knowledge and increasing the physical plant, while on the other side there have been men seeking to protect a set of definitions — a culture — that had been designed to permit man, with limited knowledge and inadequate physical plant, to deal with conditions that often seemed unchanging. To put it another way, over the past 100 years it has been, too often, as if men tried to answer Huxley's query of "What are you going to do with all these new things?" by considering only man's place in nature and without taking into account his possible place in the new environment created for him by "all these new things."

Any successful attempt to deal with the query must take the nature of the new things and environment into account. There are certain characteristics of the system of ideas, energies, and machines that must be understood at the beginning. First, the system has a rapid rate of internal change. Second, it tends toward uniformity in its products and effects. Third, it tends toward repetition and division in its procedure. Fourth, it tends to increase in mass and speed. All of these tendencies carried to extremes, as the inertia in the system seeks to carry them, produce conditions that are beyond the human powers of accommodation. Taken together, the rate of change, the uniformity, the repetition, the mass and tempo create an environment to which just physiologically human beings cannot easily adjust. People become, in the modern catchword, alienated. And the sense of alienation is increased by the fact that the system, while it may have an intellectual and empirical integrity — it works — has not moral integrity, no apparent purpose beyond effective operations.

The effort to create, consciously, a new kind of culture to contain and govern a system of this sort must take these characteristics into account as a first principle. It must devise ways to capitalize on the virtues in these characteristics and redress the defects. It must, for instance, seek to introduce the possibility of idiosyncrasy within the uniformity, of variation within the necessary repetition, of smallness within the economies of scale, of modulation within the prevailing tempo. As a set of propositions this may appear to have no more than the attractive resonance of poetic affirmation. But they can be given more substance through specific instance.

Oddly the Armed Forces have offered modest examples of how to begin to proceed. They have the nuclear weapon that has fulfilled the exaggerated extreme toward which the system always tends — the capability of total destruction. But for practical purposes they have created around this extreme a whole arsenal of carefully graded instruments of limited destruction — old-fashioned armaments of lesser power and new weapons of modulated nuclear energy.

What can be done to destroy can also be done in the interests of construction. Take speed. We have now the jet and soon will have the supersonic transport engines. But this does not mean that we must of necessity careen about on our lawful occasions always at hair-raising speeds. We can seek to fit the varying tempos of transport more nicely to the human needs or desires in a variety of these occasions. Take repetition and division. In our modern factories the energies and imagination of the largest part of our population are fragmented and locked up in meaningless reiteration of numberless small procedures that cannot easily be connected with numberless other small procedures. There are obvious economies in such methods, but this does not mean that some ingenuity in design could not produce more humanly interesting manufacturing solutions that would, in time, turn out to economic advantage at least in the stores of human energy and imagination they would make available. Take mass. Whether in cities, corporations, apartment houses, universities, or schools, the system encourages and supports the steady increase in scale. Whether it is in an overcrowded airport, along a mile-long assembly line, or in a classroom seating 2000 and wired for sound, there is the sense of everything so big that there is no definable place for a person to fit in to get the kind of attention he needs. There is no reason why a decentralizing energy cannot be introduced to reduce the scale of local units to habitable proportions.

The list of suggestive examples might be indefinitely extended. The point, however, would remain the same. The method of invention developed in the nineteenth century to produce the flood of new goods required for the material build-up of that time needs extension and redirection to meet the different needs of the twentieth and twenty-first centuries. As a beginning, within the mechanical realm invention must meet not so much — or not just — the claims of technical or financial requirements but the claims of human interest as well. The resourcefulness and understanding of engineers may well have to be enlarged if they are to build machinery that is not only efficient but satisfying — even fun — to work with.

Probably the machine that took human beings most into account and was the most fun to work with was the first serious prime mover, the steam engine. It was big enough to be imposing and small enough to be comprehended. The principal working parts were sufficiently exposed, not only to get at but to give a feeling for what was going on. It made perfectly wonderful noises and produced stimulating displays of smoke, steam, and fire. It moved at majestic tempos and gave off a sense of mass and permanence "untroubled by the dictates of cost-accountants anxious to replace [it] with something profitably new." One such engine did its work at Crofton in England, and still after one hundred and fifty years "the contemplative pauses of its mighty action was something that moved the heart. . . ."

In addition, and more important, a man had to work hard to keep it going. Gauges to study, valves to open, fires to bank, belts to tighten, levers to pull, delicate balances to maintain throughout the system. "From coupler-flange to spindle-guide/I see Thy Hand, O God./Predestination in the stride/O' yon connectin'-rod." Small wonder the engine moved the heart and entered the imagination and came down in song and story. Working at such a mechanism extended and sharpened a man's powers rather than withered them away.

It was of course a primitive system developed in man's first days of industrial knowledge. The thought here is not to return to man's first days. It is simply to suggest that effects first produced in innocence may still today, with systems infinitely more sophisticated, be achieved by design. In general this cannot so easily be done with the separate pieces of modern machinery — save in the realm of consumers goods, where more consideration for the existence and nature of the consumer would certainly help — but it could be done, more imaginatively, in the design and arrangement of the several pieces of machinery in the systems that surround us all in factories and elsewhere.

The most important kind of invention for the future lies not, however, as in the nineteenth century, within the mechanical realm, but in another area: the way we are to deal with all the new conditions produced by the new machines and ideas. In the first essay in this book the proposal is made that in view of all the new conditions and the prospect of many more to come we should seek to become an adaptive society, detached from allegiances to specific products or procedures which will change; committed instead to engagement in the process of living — that is, in the present age, to the process of rapid change itself. The proposition has a certain appeal: it was fortified by a substructure of the findings of psychology and bolstered with dignifying historical references to Elizabethan England and the Italian Renaissance. But like the preacher's injunction to "Do Good, Do Good, Do Good," it lacks something in specificity. It also somewhat begs the main question: Is adaptation by itself enough without some power to select, from possible changes, the most desirable changes to adapt to? To put it another way — in seeking merely to accommodate easily to whatever turns up in the way of new machinery or new ideas, will an adaptive society lose all meaning except survival and can it, in fact, survive very long by pursuing this sole objective?

The answer given to these questions fifteen years later is that it is not enough to be simply an adaptive society in a time of great change. Means must be discovered for society to keep charge over its own nature and direction. Three things appear necessary to achieve this end. First, the members of a society must feel that they are participating in the way affairs are ordered, that they have the power of choice. Second, to make this sense of participation and choosing real, the members of the society must have available the kind of evidence required to make judgement among possible alternatives. Third, they must have, beyond the evidence supplied for any particular case, a sense of a more general point of purpose that would serve as a governing context into which the particular judgements might be fitted.

The tendency in the present situation is to reduce the influence of these requirements. The inertia in the massive system of machinery, energy, and ideas tends, as inertia always does, toward "uniform motion in a straight line." The great agencies public and private — governments or corporations — that have been created to cope with this inertia tend equally toward the creation of uniform effects, decisions taken by others, and at a distance and applied across the board. The problem, first, is how to maintain within such a system a suitable array of alternatives for members of the society to consider before decision is taken and how to produce sufficient evidence for reasonable choice among the alternatives.

Scattered through the preceding chapters of this book there is fragmentary evidence that may suggest a possible approach to the solution of this problem. In the historical situations described, certain men, in each case, were confronted by a new kind of machinery or a new mechanical process — a new way to fire a gun, a new way to make steel, a new kind of ship, and so forth. In each case the men resorted to what may be called an experimental demonstration. H.M.S. *Terrible* under the direction of Sir Percy Scott became for two years a kind of floating laboratory of naval ordnance. At Wyandotte and Troy the converters were quite consciously built on an experimental basis. The Secretary of the Navy asked three different men to try their hand at designing the vessels that might fulfill the novel concept of Isherwood. The philanthropist used two city blocks in New York as an experiment to demonstrate the effectiveness of the pasteurizer. C. P. Snow, disturbed by the implications of the computer as a decision-maker, proposed to put it through a set of experimental exercises.

In each case when confronted by mechanical novelty men made the effort to provide interested members of the immediate community with the opportunity to get evidence on how the new instrument worked, to assess its meaning in the community, and to determine for themselves the acceptance, rejection, or modification of the new condition offered them.

The suggestion here, at a time when we are all involved in technological change and are therefore all interested and affected members of the community, the suggestion here is to introduce a massive expansion of this process throughout all parts of the society, to create the mood and means that will enable the members of the society to explore new instruments and new procedures by designed experiments while pondering alternatives and reserving judgement until the returns are in. In all the areas of difficulty and doubt — transport, the organization of cities, the control of traffic, the intelligent, indeed the loving care of the sick, the process of education, the structure of existing institutions, the means of transport, and the like — in all these areas the development of a series of small experiments, with the means available for observing the evidence produced and analyzing the results, would produce a set of alternative solutions and the data necessary both for fuller understanding of the nature of new situations and for intelligent selection among alternatives.

Take some new machinery, the elements, for instance, in the developing technology of what is now called the "knowledge industry." This technology includes not only the famous devices like computers and teaching machines and language laboratories but also an arsenal of new devices for rapid duplication of material, for taking in old and new kinds of information, for hearing, for seeing, and so forth. The implications of this rapidly developing technology are by no means fully understood; the effects it might have on what may be studied, or how it might be studied, on school architecture, on classroom size, on the learning process, and so forth, are at best dimly discerned. As things stand now the pieces of this technology make their way into the world of education, setting their own limits on the educational process. The use of the machinery, its fuller development, is in considerable part determined by what electrical engineers, chemists, manufacturers believe is technically possible and economically feasible.

A series of small experiments conducted by the men who made the machinery, by teachers, by scholars interested in developing new kinds of materials in their fields, by psychologists anxious to find out what does happen in the learning process, and by school children in classes all working together would in time produce a far fuller understanding than now exists of the potentials in the machinery, of the parts to be added and subtracted, of what further machines and procedures were necessary, of all those implications — intellectual, social, architectural, pedagogical — that are invested in this extraordinary new array of technology.

The conditions governing such experiments, whether conducted with new machinery, new building designs, new manufacturing systems, new ideas for community organization, are relatively simple and much the same in each case. Each experiment should be small enough in scale and sufficiently detached from existing practice so that the continuing state of things would not be disrupted. In addition, in each area of interest there should be not one, but a good many different experiments of differing design so that a suitable array of alternative solutions could be offered. There should also be enough of them in all parts of our life to demonstrate that the society is proceeding, as a whole, in the mood of experiment. It would, for instance, be useful to have the possibility of experiment so built into our habits of mind that provision could be made for it as a standard item — like overhead or depreciation charges — in operating budgets and grants.

The creation and preservation of this experimental mood may in itself be of the first importance. It suggests that members of the society can have a direct part in the decisions affecting the shape of the society; by offering the possibility of reasoned change, it may measurably reduce the natural human resistance to changes not fully understood. This has been from the very beginning an experimental society. By virtue of geography, exploration, immigration, the tradition of free inquiry, the nature of the federal political system, and the episodic nature of the developing technology, it has always had in the past a continuous array of novel proposals to choose from. The opportunity of free choice within a broad band of alternatives is in fact the essence of the democratic process. But at present it appears that as the weight and mass of the technical system increase, the opportunity for the whole society to renew itself by natural processes of conscious selection is sadly diminished. On the one hand there is the tendency to concentrate the power of decision in those few who seek to manage the system, and on the other hand there is the tendency in the system itself to proceed toward whatever is economically and technically possible without regard to other considerations of interest and value. The point of the experimental society is to redress those tendencies by consciously supplying a series of reasonable alternatives, varied solutions, for the whole society to think over as it makes up its mind about what it would like to do and, equally important, to be.

In the matter of making up the mind, the data in any particular case are a necessary but not a sufficient condition. Beyond the evidence, as was said earlier, a sense of further point or purpose is required, a kind of summary context within which judgements in particular cases can fit and make a contribution to the whole. What kind of grand design may there be for all this experiment, or, as Huxley put it, what are you going to do with all these things? Can there be a governing synthesis so that all goods and services may be brought nicely to support some Platonic ideal, a concept of the greater glory of God, some scheme of empire, some received doctrine, discernible at least in outline, of Progress or of an increasing purpose working through the world.

Maybe, in time, there will be found some grand design to satisfy all the data in the disturbing new world, settled doctrine from which all men can take their bearings. But that time is hardly now. For at least a season something that may seem more modest appears more appropriate. One of the purposes of the experimental society should obviously be to find out the potential and limit of the new machines and systems, but the larger context for the experiment is to find out, before things go too far, how men respond to the new conditions. The successful organizing principle for the new things is not so likely to be found in a search for how all these things may be fitted together to fufill the claims of logic, or economics, or physical laws, or mechanical elegance, but in a search for how they may be fitted together in support of man himself. Not political man, or forgotten man, or economic man, or statistical man, but the whole bundle appearing as single man.

Concerning the nature of the bundle we already know a good deal. Since the beginning, poets, theologians, philosophers, historians, prophets, and dreamers have worked to provide approximations of the immense range of a human being — so like a mole at times, at times fretted with golden fire. And in recent days the sciences, from chemistry at one end to sociology at the other, have been closing in on the quarry by more systematic observation and analysis. In the future as in the past the work of investigation and description will of course go on, and it will help. It will help more perhaps if the poets and philosophers and the scholars who serve as custodians of poetry, philosophy, and history recover some confidence in themselves and their subjects — and think, perhaps, a little less about the mole part and a little more about the golden fire. There is, after all, a kind of grandeur in the topic that need not necessarily elude all the skills of scholarship. And it will help too, no doubt, if the men of science and measurement do not believe that they can find out every single thing about the topic by their careful means. Nonetheless it all helps.

But it will not serve without further fortification. Such fortification may come by putting men and their responses at the controlling center of all the experiments with the new machines and procedures. Experience suggests that this can ordinarily be done now only by calculation and design, as in experiment. For instance, enough is now known about men to know that to live in a slum is a bad thing, but the kind of better thing built to replace a slum is, ordinarily, determined more by the state of the art in bricks, mortar, and structural steel than by any very precise feeling for what a better way of living might really be. For further instance, enough is known about men to know that ignorance is a bad thing, and so to reduce ignorance and multiply the number of informed students, a good knowledge-man is wired for sound without much discrimination in the matter of what kinds of knowledge can be learned through loud-speakers and without much thought about what damage may be done to students by this process of education misapplied.

So the proposal is to start all experiments with man as the great criterion. Put him in ten, twenty, thirty city blocks of different architectural design, social structure, transport facilities, merchandising processes, and let him live for a while to see where he lives best before starting the bulldozers of urban renewal. Put in 30, 500, 1000 schoolrooms of different architectural design, subject matter, teaching materials, and procedures to see where he learns most before one decides, as it is said we must, if there has to be a national curriculum, and if so of what kind. Put him in any number one wishes of mechanical systems — for transport, communication, making things, and so forth — to discover where he breaks down and where he thrives.

The object of all these exercises is, as has been already tiresomely repeated, to take the measure, a little more closely than heretofore, of what man is in the new environment he has created for himself and to give him the evidence necessary to modify, limit, and organize the developing environment so that he may extend his own range within it.

Index

Index